Designing for Transportation Management and Operations

A PRIMER

U.S. Department of Transportation
Federal Highway Administration

February 2013

Notice

This document is disseminated under the sponsorship of the U.S. Department of Transportation in the interest of information exchange. The U.S. Government assumes no liability for use of the information contained in this document. This report does not constitute a standard, specification, or regulation.

The United States Government does not endorse products or manufacturers. Trademarks or manufacturers' names appear herein only because they are considered essential to the objective of this document.

Quality Assurance Statement

The Federal Highway Administration (FHWA) provides high quality information to serve Government, industry, and the public in a manner that promotes public understanding. Standards and policies are used to ensure and maximize the quality, objectivity, utility, and integrity of its information. FHWA periodically reviews quality issues and adjusts its programs and processes to ensure continuous quality improvement.

Technical Report Documentation Page

1. Report No. FHWA-HOP-13-013	2. Government Accession No.	3. Recipient's Catalog No.
4. Title and Subtitle Designing for Transportation Management and Operations: A Primer		5. Report Date February 2013
		6. Performing Organization Code
7. Author(s) (in alphabetical order) Jennifer Atkinson (SAIC), Jocelyn Bauer (SAIC), Kevin Hunt (GF), Keith Mullins (GF), Matthew Myers (SAIC), Eric Rensel (GF), Myron Swisher (SAIC), Robert Taylor (GF)		8. Performing Organization Report No.
9. Performing Organization Name and Address Science Applications International Corporation (SAIC) 8301 Greensboro Drive McLean, VA 22102 Gannett Fleming, Inc. 1515 Market Street, Suite 2020 Philadelphia, PA 19102-1917		10. Work Unit No. (TRAIS)
		11. Contract or Grant No. DTFH61-06-D-00005
12. Sponsoring Agency Name and Address United States Department of Transportation Federal Highway Administration 1200 New Jersey Ave., SE Washington, DC 20590		13. Type of Report and Period Covered
		14. Sponsoring Agency Code HOP
15. Supplementary Notes Mr. Jim Hunt, Federal Highway Administration, GTM		
16. Abstract This primer is focused on the collaborative and systematic consideration of management and operations during transportation project design and development. This is termed "designing for operations." Effectively designing for operations involves the development and application of design policies, procedures, and strategies that support transportation management and operations. The consideration of operations needs during the design process requires transportation design professionals to work closely with those with expertise in transportation operations, intelligent transportation and transportation technology staff, planning, transit, freight, traffic incident management, and other practitioners from multiple agencies to fully identify, prioritize, and incorporate operations needs into the infrastructure design. This primer introduces the concept for designing for operations and describes tools or institutional approaches to assist transportation agencies in considering operations in their design procedures as well as pointing out some specific design considerations for various operations strategies.		
17. Key Words Designing for operations, management and operations, intelligent transportation systems, planning for operations, design standards, project development.	18. Distribution Statement No restrictions.	
19. Security Clasif. (of this report) Unclassified	20. Security Clasif. (of this page) Unclassified	21. No. of Pages 52	21. Price N/A

Form DOT F 1700.7 (8-72)   Reproduction of completed page authorized

Acknowledgements

The development of this primer greatly benefited from the contributions of practitioners from State and local departments of transportation, metropolitan planning organizations, and transit agencies. Federal Highway Administration and the authors acknowledge the individuals who provided input through peer exchanges and interviews:

Ted Bailey, Washington State Department of Transportation

Morgan Balogh, Washington State Department of Transportation

John L. Benda, Illinois Tollway

Elizabeth Birriel, Florida Department of Transportation

Rashmi Brewer, Minnesota Department of Transportation

Saahir Brewington, American Public Transportation Association

Grady Carrick, Florida Highway Patrol

Yi-Chang Chiu, University of Arizona

Michael Curtit, Missouri Department of Transportation

Gene Donaldson, Delaware Department of Transportation

Kay Fitzpatrick, Texas Transportation Institute

Chris Francis, Virginia Department of Transportation

Bernie Guevara, Colorado Department of Transportation

Brian Hoeft, Regional Transportation Commission of Southern Nevada – FAST

Denise Inda, Nevada Department of Transportation

Chris King, Delaware Valley Regional Planning Commission

Walter Kraft, Vanasse Hangen Brustlin, Inc.

Monica Kress, California Department of Transportation

Beverly Kuhn, Texas Transportation Institute

Lacy Love, American Association of State Highway and Transportation Officials

Alvin Marquess, Maryland State Highway Administration

David Millar, Fehr and Peers

Richard Montanez, City of Philadelphia

Yuko Nakanishi, Nakanishi Research and Consulting

Thomas Notbohm, Wisconsin Department of Transportation

Jim Rosenow, Minnesota Department of Transportation

Faisal Saleem, Maricopa County Department of Transportation

Robert L. Trachy Jr., Virginia Department of Transportation

Table of Contents

1 DESIGNING FOR OPERATIONS 2
    1.1 Introduction to Designing for Operations 2
    1.2 Management and Operations Overview 4
    1.3 Why Integrate Operations into Design? 5
    1.4 Primer Audience and Overview 5
    1.5 Examples of Effective Designing for Operations Practices 6
    1.6 Limitations in Current Design Practices 9
        1.6.1 Institutional Challenges 10
        1.6.2 Fiscal Impacts of Designing for Operations 10
        1.6.3 Understanding Management & Operations Needs 11

2 PUTTING IT INTO ACTION – POLICIES AND PROCEDURES 12
    2.1 Institutional Policies 12
    2.2 Agency Structure 14
        2.2.1 Culture and Leadership 14
        2.2.2 Organization and Staffing 14
    2.3 Linking Planning and Designing for Operations 15
    2.4 Project Development Process 17
        2.4.1 Scoping & Financing Stage 18
        2.4.2 Preliminary Design Stage 18
        2.4.3 Final Design Stage 19
        2.4.4 Examples of Designing for Operations in Project Development 19
    2.5 Systems Engineering 21
    2.6 Design Standards and Checklists 22
        2.6.1 One Stop Manual for Engineering and Other Technical Information ... 22
        2.6.2 Operational Review and Sign-off of Standard Plans and Specifications ... 24
        2.6.3 Operations Audits and Review Team 25

3 DESIGN CONSIDERATIONS FOR SPECIFIC TYPES OF
    OPERATIONS STRATEGIES 26
    3.1 Freeway Management 26
    3.2 Arterial Management 28
        3.2.1 Cooperation of Municipalities 29
        3.2.2 Managing Access for All Modes 29
        3.2.3 Monitoring and Actively Managing Traffic Conditions and Intersection
               Signalization 30
    3.3 Active Traffic Management 31
    3.4 Managed Lanes 33
    3.5 Transit 35
    3.6 Work Zone Management 37
    3.7 Traffic Incident Management 38
    3.8 Security 40
    3.9 Freight Operations 41
    3.10 Maintenance 42

4 A WAY FORWARD 44

1 Designing for Operations

1.1 INTRODUCTION TO DESIGNING FOR OPERATIONS

Transportation agencies across the United States are looking for ways to provide safe, efficient, and reliable travel across modes and jurisdictions under increasingly constrained fiscal environments. The public, business leaders, and elected officials want reliable goods movement, timely and accurate traveler information, safe and quick incident clearance, and greater options in transportation modes, routes, and services. To address the need for mobility, safety, and security, many transportation agencies have integrated management and operations into their set of solutions. Management and operations strategies can often improve transportation system performance significantly and be deployed more quickly and more cost-effectively than traditional capacity expansion projects. Management and operations (M&O) strategies focus on getting the most efficient and safest use out of existing or planned infrastructure through activities such as traffic incident management, traveler information dissemination, traffic signal coordination, and work zone management. M&O strategies are funded and implemented as stand-alone projects or combined with larger projects such as highway reconstruction.

Figure 1. Locations of variable message signs should be selected for maximum effectiveness such as in areas of frequent hazardous weather or traffic incidents and prior to traveler decision points. (Source: Nevada Department of Transportation)

The effective management and operation of the transportation system often requires traditional infrastructure (e.g., roadways and other civil infrastructure) to be designed to support M&O strategies. This includes roadway design for freeways and arterials, transit system design, as well as strategic integration of intelligent transportation systems (ITS) on roadways and rail systems. For example, a high volume highway segment with a full-depth shoulder sufficient to support traffic is needed for bus-on-shoulders, an M&O strategy in which only public buses may use the shoulder to minimize delay during peak congestion periods. Additionally, the installation of variable message signs (VMS) in locations prior to significant route or modal decision points for travelers or common incident areas supports relevant, actionable traveler information to the public. Other examples of roadway design treatments that are important for improving the management and operation of the facility include:

- Median crossovers, which allow for incident responders to quickly access the opposite side of the road;
- Crash investigation sites, which reduce impacts associated collecting incident information;
- Snow fences, which reduce blowing snow and drifts on the road; and
- Emergency access between interchanges, which decrease response time to incidents; and
- Bus turnouts, which ease arterial congestion.

Traditionally, the needs of M&O strategies have not been fully considered in roadway infrastructure design, including major projects such as road, bridge, and tunnel construction, roadway expansion or extension, bridge restoration, tunnel rehabilitation, and re-paving. Roadway design processes take into consideration some aspects of how the facility will operate by considering peak hour design traffic volumes, expected truck volumes, signing, striping and pavement marking needs, desired design level of service, and design speed. However, critical M&O considerations that support the broad array of strategies listed in Section 1.2 are often addressed in an ad-hoc manner or are given insufficient consideration during project development. This means that the operations needs of the system are either not sufficiently addressed or agencies must retrofit roadways after they are constructed or reconstructed. This latter approach is usually less effective and more expensive

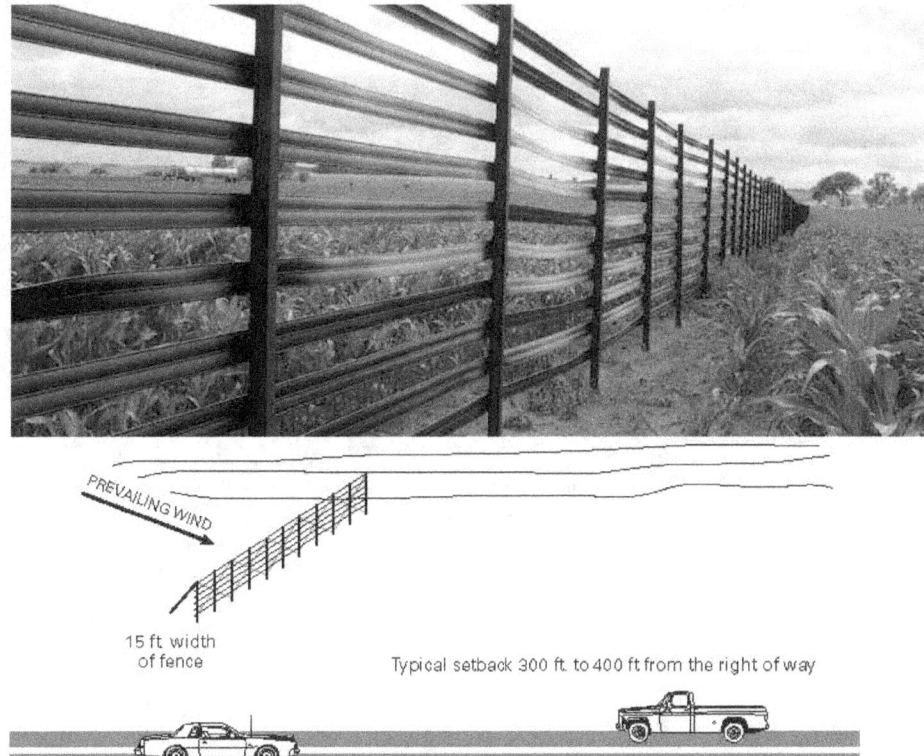

Figure 2. An example of a design treatment to improve the operations of a roadway is snow fences that reduce blowing snow and drifts on the road. Shown are the simple design guidance and an image of the corresponding implementation in Minnesota. (Photo Source: Dan Gullickson, MnDOT, Diagram Source: MnDOT)

in terms of construction, user, and right-of-way costs than including operational needs in the original design. At times, retrofitting the roadway for operations forces the need for design exceptions that may raise safety concerns. These are issues that typically can be eliminated if operations is considered during the design and preliminary engineering stages.

This primer is focused on designing for operations; i.e., the collaborative and systematic consideration of management and operations during transportation project design and development. Effectively designing for operations involves the development and application of design policies, procedures, and strategies that support transportation management and operations. Considering operations needs during the design process requires transportation design professionals to work closely with those who have expertise in transportation operations, intelligent transportation, and transportation technology. Design professionals should also anticipate working with practitioners from planning, transit, freight, and TIM as well as staff from other agencies in order to fully identify, prioritize, and incorporate operations needs into the infrastructure design. There are several entry points for integrating M&O strategies into the traditional project design process, as described in this primer. Designing for operations is typically reflected in increased or formalized collaboration between designers and operators and the development of design guidelines and procedures that reflect a broad range of operational considerations.

Successful integration of M&O considerations into the design process means that:

- Roadway and transit system infrastructure is designed to facilitate the needs of day-to-day system management and operations and meet transportation system performance targets for efficiency, reliability, travel options, and safety.
- ITS deployments are designed using systems engineering, and existing and future operational uses of ITS are incorporated into transportation facility design.
- Operational strategies are considered as credible alternatives to infrastructure expansion during project design. The relatively low-cost, high-impact, and flexible nature (i.e., scalable to changing demands) of M&O strategies makes them attractive deployment options.

There are regions and States across the United States that are moving forward with the incorporation of operations needs into project design because they recognize the benefits in terms of cost savings and system performance. For example, the published policies and procedures of the Regional Transportation Commission (RTC) of Southern Nevada require all projects to be designed to the standards of the Regional Intelligent Transportation Systems Architecture adopted by the RTC.[1] Additionally, copies of all project plans must be provided to the Freeway and Arterial System of Transportation (FAST) organization for review. The policies and procedures also require the consideration of raised medians to reduce left turn conflicts and pedestrian refuge during project design. Also, installing conduit should be considered during project construction if traffic signals are anticipated in the future. Another example is from the Delaware Department of Transportation where there is a review of M&O requirements in each design phase of every capital transportation project.

The congestion management process (CMP) has also served as a motivator for designing for operations. For example, the Delaware Valley Regional Planning Commission (DVRPC) requires as part of its CMP that any project that adds major capacity for single-occupancy vehicles (SOV) includes supplemental strategies to reduce congestion and get the most from infrastructure investments. The final engineering for a major SOV capacity adding project must include a list of supplemental strategies to be included in the transportation improvement program (TIP) for funding. These supplemental strategies — such as traffic signal improvements, signal preemption for emergency, park-and-ride lots, and engineering strategies to improve traffic circulation — work to improve the overall management and operation of the facility.[2]

Other examples of designing for operations practices can be found in Section 1.5.

Figure 3. Express high occupancy/toll lanes with variable pricing is an M&O strategy increasingly used in the United States.
(Source: SAIC)

1.2 MANAGEMENT AND OPERATIONS OVERVIEW

Systematic consideration of M&O strategies during the design process is at the core of designing for operations. Transportation systems management and operations is defined by the legislation "Moving Ahead for Progress in the 21st Century" (MAP-21) as the use of "integrated strategies to optimize the performance of existing infrastructure through the implementation of multimodal and intermodal, cross-jurisdictional systems, services, and projects designed to preserve capacity and improve the security, safety, and reliability of the transportation system."[3] M&O strategies encompass many activities, such as:

- Traffic incident management.
- Traffic detection and surveillance.
- Corridor, freeway, and arterial management.
- Active transportation and demand management.
- Work zone management.
- Road weather management.
- Emergency management.
- Traveler information services.

- Congestion pricing.
- Parking management.
- Automated enforcement.
- Traffic control.
- Commercial vehicle operations.
- Freight management.
- Coordination of highway, rail, transit, bicycle, and pedestrian operations.

---

[1] Regional Transportation Commission of Southern Nevada, *Policy and Procedures Manual*, Revised September 13, 2012. Available at: http://www.rtcsnv.com/wp-content/uploads/2012/03/RTC-Policies-Procedures.pdf.

[2] Delaware Valley Regional Planning Commission, *Overview of the 2011 Congestion Management Process (CMP) Report*, May 2011. Available at: http://www.dvrpc.org/asp/pubs/publicationabstract.asp?pub_id=11042A.

[3] MAP-21, SEC. 1103. Definitions. http://www.fhwa.dot.gov/map21/legislation.cfm.

Management and operations also includes the regional coordination required for implementing operational investments such as communications networks and traffic incident management in an integrated or interoperable manner.

Successful M&O practices positively impact mobility, accessibility, safety, reliability, community life, economic vitality, and environmental quality and help transportation agencies meet their customers' needs. In addition, many agencies have found that the benefits of M&O strategies can significantly outweigh the costs (versus traditional strategies). Proactive management of transportation systems in real-time and at all hours of the day not only represents the future of operations but is essential to responding effectively to variable traffic conditions caused by events such as incidents, work zones, and weather effects.

1.3 WHY INTEGRATE OPERATIONS INTO DESIGN?

To maximize the safety, reliability, and efficiency of the transportation system, it is crucial that roadways, bridges, and transit infrastructure be designed to better manage demand and respond to incidents and other events. Designing for operations improves the integration of operational considerations throughout the transportation project development lifecycle, resulting in better resource utilization, improved maintenance and asset management practices through enhanced collaboration, and effectively designed and deployed infrastructure improvements. Some advantages of incorporating operations into traditional design processes include:

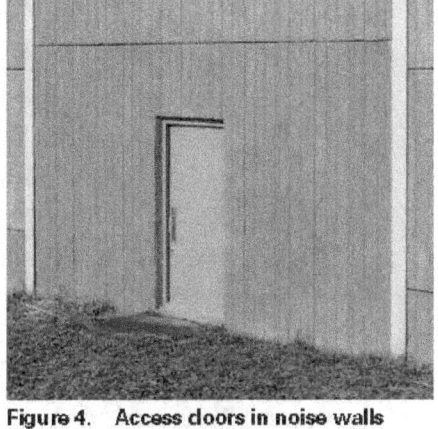

Figure 4. Access doors in noise walls like this one on Virginia State Route 267 can decrease incident response time. (Source: SAIC)

- Increasing the benefits derived from a given infrastructure investment.
- Designing a safer facility for users, emergency responders, maintenance staff, and other operators.
- Designing for future work zones so that road users experience less interruption.
- Reducing the costs for future operational and ITS deployments.
- Reducing congestion and improving travel time reliability.

Applying the concepts of designing for operations in a cohesive and standardized manner decreases long-term costs, saves contract and staff labor time, and can improve operational performance of the transportation system. For example:

- Installing conduit during major roadway or bridge reconstruction projects on a corridor can reduce the future communications systems costs of the corridor's freeway management system.
- Including a full-depth shoulder with sufficient design to support traffic in a reconstruction project can allow the shoulder to be utilized as part of an alternative capacity expansion concept such as allowing bus use of highway shoulders.
- Locating an access door in a noise wall at a critical location can decrease the response time to a major incident on a freeway.

1.4 PRIMER AUDIENCE AND OVERVIEW

This primer introduces the concept of designing for operations, describes tools and institutional approaches to assist transportation agencies in considering operations in their design procedures, and points out some specific design considerations for various operations strategies. The tools and approaches to aid in designing for operations may include checklists for designers to reference operational considerations, formation of a technical advisement committee with operations expertise, or agency policies that instruct designers on how to incorporate operational elements within the project development process. These will benefit multiple practitioner groups, including planners, project designers, scoping engineers, maintenance and traffic managers, and contract development personnel.

The primer is written for several primary audiences:

- Roadway designers.
- Transportation planners.
- Transportation operations professionals.
- Transportation agency managers.

The primer is organized into four sections:

Chapter 1. Designing for Operations (the current chapter) – This chapter introduces the term "designing for operations" and explains its value and purpose. It orients the reader to this primer and provides examples of successful designing for operations practices being implemented by State departments of transportation (DOTs) and metropolitan planning organizations (MPOs). Lastly, it identifies limitations in current design practices that must be overcome in order to enable effective designing for operations.

Chapter 2. Putting it into Action - Policies and Procedures – This chapter defines the key actions agencies can take to better incorporate operational considerations into the design process. It identifies opportunities to implement policies and procedures and to encourage communication and collaboration between operations, design, and planning disciplines. It also provides general approaches for agencies to incorporate operations into the transportation design process.

Chapter 3. Design Considerations for Specific Types of Operations Strategies – This chapter provides a toolbox of design considerations to support M&O strategies.

Chapter 4. A Way Forward – This chapter ties together the major themes of the primer and provides a concise summary of the primary lessons for the reader.

Who Should Read This Document and Why?

- **Roadway designers** will gain perspective regarding the types of operational strategies that are available and how features that support those strategies can be integrated into traditional design.
- **Transportation operations** professionals will be engaged to ensure that operational strategies are included in the design process where appropriate.
- **Transportation planners** will be equipped to better identify project relationships and synergies (such as those between infrastructure and operations) that can help improve the long range planning and transportation improvement program development processes.
- **Transportation agency managers** will understand the benefits that can be derived from placing a greater emphasis on operations throughout the project lifecycle.

1.5 EXAMPLES OF EFFECTIVE DESIGNING FOR OPERATIONS PRACTICES

Across the United States, there are several agencies that have adopted effective approaches for integrating operations considerations into the design of transportation infrastructure. The examples highlighted below illustrate a range of practices, from the consideration of operational improvements as part of a project design prior to programming the project, to the use of ITS design manuals and guidelines. A few of the examples showcase collaboration between designers, operations experts, and professionals with other areas of expertise to ensure that project designs take into account a broad range of needs.

The California Department of Transportation (Caltrans) and its local partners develop Corridor System Management Plans (CSMP) for those corridors that are heavily congested. Through these efforts, M&O strategies are routinely identified and included as a package of improvements for the corridor.[4] The CSMP identifies bottlenecks, which are defined as "localized sections of highway where traffic experiences reduced speeds and delays due to recurring operational conditions or nonrecurring traffic-influencing events," and looks for opportunities to address the bottlenecks. Bottleneck reduction strategies include several M&O strategies such as traffic signal re-timing, access management, and providing traffic diversion information. As part of this effort, Caltrans may conduct a localized bottleneck reduction audit by reviewing traditional large-scale corridor studies in order to identify opportunities to deploy low cost bottleneck reduction strategies within a package of improvements.

In 2012, Portland Metro, the City of Portland, Portland State University, and Metro's regional partners developed *The Portland Multimodal Arterial Performance Management Regional Concept of Transportation Operations* (RCTO) to provide regional guidance for collecting automated multimodal performance measures on arterial roadways.[5] The RCTO is the Portland region's next step in advancing performance-based transportation planning and investment decision-making. The guidance includes detector technology options, design considerations for installation of data collection technology, and intersection diagrams depicting where detection is needed. The RCTO recommends updates to the Oregon DOT and local agency design standards. As new projects arise, Metro is encouraging the installation of necessary field equipment and communications to collect all eligible performance measures at given locations.

Figure 5. Diagram from *Portland Multimodal Arterial Performance Management Implementation Guidance* depicting typical locations for collecting multimodal arterial data.[6]

---

[4] U.S. DOT FHWA, *An Agency Guide on Overcoming Unique Challenges to Localized Congestion Reduction Projects*, Sep 2011. Available at: http://ops.fhwa.dot.gov/publications/fhwahop11034/ch3.htm.

[5] Metro, *Portland Multimodal Arterial Performance Management Implementation Guidance*, Unpublished Draft, 2012.

[6] Ibid.

The Florida DOT's Transportation Systems Management and Operations (TSM&O) Program[7] is a prime example of a program that formalizes designing for operations practices through stakeholder collaboration. The TSM&O Program is implemented throughout all departments, including design, where operational aspects of a facility are to be designed and built during construction for easier long-term operations and maintenance of the facility.[8] The program's focus is measuring performance, actively managing the multimodal transportation network, and delivering positive safety and mobility outcomes to the traveling public. To do this, TSM&O relies on collaborative relationships between partners. The TSM&O partners are comprised of public transportation agencies who serve together as one cohesive entity to make cost-effective investment decisions. This cohesiveness serves to improve communications, coordination, and collaboration amongst transportation partners with diverse perspectives, leading to more designs that consider operations.

The Florida DOT District 4 has established a project development and management structure that formally includes input from operational agencies, such as signal and transit departments. The objective is to reflect operational elements in project design and operational strategies that are consistent with the current and planned infrastructure. The overall process makes it clear to the District how its operational requirements are being reflected in project development.

The Washington State DOT (WSDOT) has formed a System Operations and Management (SOM) Committee, which is comprised of a cross-cutting group of program managers from the DOT's regions and headquarters. Currently, the group is working to shape the design process to be more flexible for implementing creative operational solutions. Each member represents an interest area such as operations, system performance, design, planning, tolling, program management, multi-modal, commercial vehicles, and travel information. The group is tasked with identifying and solving the many evolving issues that arise as the agency moves toward being more operations focused. The committee's ultimate goal is to improve capacity and travel reliability *without* traditional levels of infrastructure investment. WSDOT also provides design guidelines for minor operational enhancement projects (low-cost enhancements intended to improve the safety and efficiency of the highway system) in the WSDOT Design Manual, *Minor Operational Enhancement Projects* (Chapter 1110).[9] These projects are considered as alternatives to larger, more costly traditional projects. An important characteristic of these projects is the ability to quickly develop and implement them without a cumbersome approval process. In order to achieve this, design policies and guidelines are applied in the development and approval processes in the same manner with each project. This approach is part of the DOT's statewide Minor Operational Enhancement Projects Program (referred to as the "Q" Program), which is one of the four major programs (i.e., improvement, maintenance, preservation, and traffic operations) within the WSDOT's Highway System Plan. Elements within the Q Program include: traffic operations program management, traffic operations program operations, and special advanced technology projects.[10]

The Pennsylvania DOT has integrated operational considerations into its design manual series and has developed the *Intelligent Transportation Systems Design Guide* which documents strategy selection, design variables, site selection considerations, systems engineering considerations, and site design considerations. Details are provided on which devices are appropriate based on roadway conditions as well as how ITS elements should be deployed within existing constraints, such as the cost of connection to communications and power utilities, safe access for maintenance staff, and available right-of-way. This guide provides a level of detail that allows non-ITS practitioners to incorporate ITS design elements into projects in a cohesive and consistent manner.[11]

---

[7] Florida DOT, *Transportation Systems Management & Operations*, http://www.dot.state.fl.us/trafficoperations/TSMO/TSMO-home.shtm.

[8] National Transportation Operations Coalition, Talking Operations Webinar, Presentation by Elizabeth Birriel, P.E., Florida Department of Transportation, August 2, 2011. Available at: http://ntoctalks.com/web_casts_archive.php.

[9] Washington State DOT, *WSDOT Design Manual*, June 2009. Available at: http://www.wsdot.wa.gov/publications/manuals/fulltext/M22-01/1110.pdf.

[10] Ibid., p. 1110-1.

[11] Pennsylvania Department of Transportation, PennDOT Publication 646 Intelligent Transportation Systems Design Guide, April 2011. Available at: ftp://ftp.dot.state.pa.us/public/pubsforms/Publications/Pub%20646.PDF.

It is important to note that the example practices and programs above are mainstreamed into the agencies' everyday business practices. WSDOT has a section in its design manual dedicated to the Q program, and Florida DOT has moved forward with a strategic plan and business plan for how to implement TSM&O within the existing policies and procedures of the DOT. Missouri DOT has established the use of core teams as a preferred practice, as mentioned in its *Engineering Policy Guide*.[12] Caltrans promotes localized bottleneck reduction audits at the district level to identify improvements related to bottlenecks. These programs foster the type of cross-cutting strategies necessary for designing for operations.

1.6 LIMITATIONS IN CURRENT DESIGN PRACTICES

Some transportation agencies are now placing a greater emphasis on operations during the design process; however, the methods for applying this practice are inconsistent due to gaps that exist within current project development processes and a lack of sufficient communication between designers and operators.

Many agencies at the State, regional, and local levels have embraced M&O practices and the use of ITS. The U.S. DOT ITS Deployment Tracking data suggest steady expansion of the Nation's ITS infrastructure and operational practices.[13] A significant number of States and regions are guided by formal M&O or ITS strategic plans and architectures and are making progress in planning for operations

Despite this progress, there is still a disconnect between infrastructure design and operations in many transportation agencies. Current processes for the development of roadway or bridge project plans typically involve following a set of design criteria, such as agency-specific guidelines or those published in the American Association of State Highway and Transportation Officials' (AASHTO) *A Policy on Geometric Design of Highways and Streets*.[14] These criteria require highway engineers to identify design controls, some of which are pre-determined (e.g., terrain, urban vs. rural, classification of the road), while others are project-specific. The Federal Highway Administration (FHWA) and AASHTO both note that the basis of geometric criteria should include additional factors such as cost, maintainability, safety, and traffic operations.[15]

Reasons for the disconnect between design and operations include the lack of policies that support the integration of M&O considerations into design as a routine way of doing business and the lack of involvement of internal and external facility operators during the project development process. Facility operator perspectives are critical not only for identifying necessary operations infrastructure, but for providing input into design decisions that influence ongoing maintenance and support of the deployed infrastructure and the manner in which the infrastructure will be used on a day-to-day basis. For example, agency traffic management operations personnel could provide recommendations on the placement and orientation of traffic surveillance equipment to ensure the data collected meet operational objectives and to facilitate access to the equipment for future maintenance.

In some transportation agencies, roadway designers may work in settings where there is limited collaboration with operations staff, so the resulting project delivery process does not maximize the opportunity to incorporate operations elements. The consequences of this disconnected process are often that more time and money are used to retrofit infrastructure for enhanced operations or that the transportation system does not perform to its potential level of service.

---

[12] Missouri DOT, *Engineering Policy Guide*. Available at: http://epg.modot.org/index.php?title=Main_Page.
[13] USDOT RITA, ITS Deployment Tracking Survey Results. Available at: http://www.itsdeployment.its.dot.gov/.
[14] AASHTO, *A Policy on Geometric Design of Highways and Streets*, 6th Edition, 2011. Available at: https://bookstore.transportation.org/collection_detail.aspx?ID=110.
[15] AASHTO, A Guide for Achieving Flexibility in Highway Design, 2004.

1.6.1 Institutional Challenges

Due to the major cost of providing new capacity, agencies have started to focus on operations as a tool to maximize the efficiency of transportation networks. But even as operations is beginning to hold an elevated status, operations and design are still too often disconnected in transportation agencies. When transportation operations, planning, and roadway design occur in separate silos within an organization or region, awareness and understanding of the need to design for operations among agency staff and management is limited, and opportunities to maximize the performance of infrastructure investments are lost. By integrating these worlds, the system performance advantages and cost savings of operational improvements may be leveraged. As emphasis shifts from building new roadway facilities to maximizing operations on existing roadways, agencies are instituting mechanisms for inter-departmental collaboration. Cross-cutting working groups such as traffic incident management (TIM) teams and multi-disciplinary project development teams have yielded improved coordination among stakeholders.

> Successful systematic consideration of operations in the design process typically takes place within an organizational setting in which operations is considered just as crucial as design, construction, and maintenance. Agencies that practice designing for operations well often have a formal policy related to those topics or broad-based, high-level support within the agency. The responsibility for incorporating operations into design cannot fall only on the designers; it must be an agency-driven approach.

The key to integrating planning, design, and operations is for all the players involved to understand their respective levels of focus and to develop tools and mechanisms for communication. Open and regular communication between core members of the project design team and agency management establishes how operational practices will be considered during the design process. High-level support within an agency and a collaborative approach to infrastructure are needed. The personnel who facilitate roadway operations include both transportation agency staff and outside stakeholders, such as transit operators, emergency responders, freight operators, and special event operators. By embracing input from outside stakeholders, a transportation agency can design a facility that enables management and operations, thereby enhancing safety and the overall transportation experience.

In addition to functional silos within an agency or region, other institutional challenges that limit the incorporation of operations into roadway design include a lack of policy or design standards that compel project development staff and designers to account for M&O strategies in infrastructure projects. It is important to note that design personnel at State and local DOTs typically do not act on their own without agency policies or management direction. This reinforces the importance of agency-wide policies and management support in advancing designing for operations. In addition, decisions regarding project goals, scope, and budget constraints are often made well before a project advances into design, so project development staff or supporting contractors should also be highly involved in working to integrate operations considerations into projects from their inception.

1.6.2 Fiscal Impacts of Designing for Operations

Another barrier to operations consideration in design is that when M&O strategies are proposed for inclusion into a roadway infrastructure design, the additional costs to do so may be construed as an unwarranted "accessory." When project costs expand beyond established budgets, operations features such as ITS and emergency responder facilities may be cut because they are deemed non-essential or low priority. However, not fully accounting for operations in design oftentimes results in higher long-term costs. This is especially true in cases where future operational deployments are planned. For example, it is much more cost-effective to place conduit along a corridor where future signalized intersection improvements are planned or construct full-depth shoulders for possible expansion or lane shifts than it is to install these treatments after the primary roadwork is completed. As reported in a recent letter from the U.S. Government Accountability Office regarding the "Dig Once" Executive Order, installing conduit and fiber as a standalone project can cost 15 to 33 percent more than when included in a roadway construction project.[16]

---

[16] Government Accountability Office (GAO), Planning and Flexibility Are Key to Effectively Deploying Broadband Conduit through Federal Highway Projects. Available at: http://gao.gov/assets/600/591928.pdf.

To maximize the value from an infrastructure investment, transportation agencies must consider the full range of alternatives and full life-cycle cost implications and then implement the most practical solution. Service life and ongoing maintenance costs of the investment must be added to project costs, while also factoring road user and environmental costs. By involving operators in the design process, project designs will better account for agency operations and maintenance resources and expectations. Depending on agency capital and operations funding levels and commitment to meeting equipment maintenance and replacement needs, design treatments for operations could be customized with the appropriate degree of "hardening" or reliability. Ultimately, designs should maximize the benefit-cost ratio over time, and that will likely include M&O strategies with their low cost and high value.

1.6.3 Understanding Management & Operations Needs

Designers and project development staff typically have a thorough understanding of the project development process but have had limited exposure to operational needs. Without experience in the practical application of M&O strategies, designers have no fundamental understanding of how their design may impact roadway operations. As recommended previously, building opportunities for operations, planning, project development, and design staff to regularly collaborate by removing functional silos from organizations will help to increase this understanding.

In addition, design guidance is needed. Through the iterative design process, designers apply appropriate agency design standards, policies, and practices depending on the stage to which the plans have progressed. Design practices typically balance infrastructure needs with project costs and consider finite design elements related to those established in AASHTO's Policy on Geometric Design of Highways and Streets and the *Roadside Design Guide*.[17] To date, an authoritative voice on operational elements to be considered during the design process has not been established or disseminated, which is a challenge to promoting widespread consideration of M&O strategies in design.

---

[17] AASHTO, *Roadside Design Guide, 4th Edition*, 2011.
Available at: https://bookstore.transportation.org/collection_detail.aspx?ID=105.

2 Putting It Into Action – Policies and Procedures

Designing for operations requires a change in the way transportation agencies conduct business. Mainstreaming operations considerations into the project design process requires an agency to develop specific policies and formal procedures. By doing so, agencies ensure that designing for operations will occur on a routine, consistent basis that transcends the effort of one or two individual champions. This often takes place in organizations where operations is elevated in organizational structure as well as in management priorities.

Complementing internal high-level support, transportation agencies must take a collaborative approach to designing for operations that includes personnel who facilitate roadway operations, such as transportation agency staff (e.g., maintenance operators, freeway service patrol staff), and other operators, such as TIM stakeholders and emergency response personnel.

In addition to bringing together operators and designers, an effective approach to designing for operations must also include a connection to transportation planners and the planning process. It is in the planning process where transportation goals, objectives, performance measures, and strategies, projects, or programs are identified and agreed upon by leaders in the State or region. Linking project design considerations to operations-based objectives, performance measures, and M&O strategies selected for the area's transportation plan promotes the systematic and performance-based consideration of operational strategies during project development and design.

Designing for operations also includes the development and implementation of ITS in a way that best supports the operation of the transportation system. This requires the use of the systems engineering approach, which focuses on the systematic consideration of how ITS will be used and what will be required of the system through the development of a concept of operations and system requirements early in the project development process. Section 2.5 provides more information on the systems engineering approach.

This section describe how designing for operations can be best supported through the policies and procedures of a transportation agency, including agency structure, institutional policies, planning and systems engineering, and each stage of the roadway project design process.

2.1 INSTITUTIONAL POLICIES

Agencies are beginning to embrace designing for operations by instituting policies that require designers to elicit input from operators and other stakeholders. For example, State DOT policies may stipulate that standard plans and specifications for a project must be reviewed by operations staff and approved by a team of represented stakeholders, from disciplines such as construction, design, operations, and districts. Establishing this policy at the statewide level ensures that all districts operate under uniform conditions.

It is important that these policies ensure that collaboration between designers and operators—both the users of the implemented systems as well as those responsible for maintaining the systems' intended functionality—begins before the design process actually starts. Whether it is an internal agency or outside consultant leading the design, the scope of work for the designer should be determined with management and operations in mind. Once the designer already has his task or scope, it is more difficult to incorporate operations considerations retrospectively.

Institutional policies can take the shape of establishing formal or ad hoc groups to collaborate on design and operations aspects of projects. The downside to ad hoc groups or policies is that they often lack a succession plan; if staff members responsible for the collaboration are promoted, reassigned, or retire, the benefits of the group can be lost. Formal policies that require designers to solicit input from operations staff and stakeholders more firmly establish the process and set expectations that collaboration is a requirement, not just an option.

Policies that require operations performance measures to be discussed and determined as part of the design stage can also support the integration of operations considerations into design. If designers have to consider how their design will meet their agency's established performance expectations at the system or facility levels, operations is more likely to be discussed earlier in the design process.

The following are examples of institutional policies that support the designing for operations approach:

- The Metropolitan Council, the MPO for the Minneapolis/St. Paul, Minnesota region, has a policy statement identifying operations as a high priority for the region and encouraging Minnesota DOT, counties, and cities to consider lower cost congestion mitigation and safety improvements in preservation and maintenance projects.[18]

- Caltrans has a policy that requires the consideration of high occupancy vehicle (HOV) lanes, park-and-ride facilities, and transit facilities prior to the project approval stage for several types of projects, such as capacity additions to freeways in metropolitan areas.[19] Caltrans requires the detailed design of those features at

Figure 6. Institutional policies, such as considering bus turnouts, promote operations in the design process. (Source: SAIC)

the plans, specifications, and estimate or detailed design stages. The transit facilities that are required for consideration where appropriate include bus turnouts, passenger loading areas, benches and shelters, and traffic control devices.

- The Missouri DOT has a policy for collaboration among those with different areas of expertise during project development.[20] Missouri DOT uses a "core team" concept that establishes regularly occurring project development meetings comprised of representatives from design, planning, construction, maintenance, traffic, right-of-way, public outreach, permits, and other relevant stakeholders. During these meetings, these representatives communicate the impact the current design direction will have on their respective discipline area, should the design be implemented as-is. The group works together to decide which strategies and design considerations are ultimately implemented and constructed. This ensures that the operations staff has a voice at the table.

---

[18] Metropolitan Planning Council, 2030 Transportation Policy Plan, 2010.
Available at: http://www.metrocouncil.org/Transportation/Planning/2030-Transportation-Policy-Plan.aspx.

[19] Caltrans Office of Project Development Procedures, *How Caltrans Builds Projects*, August 2011.
Available at: http://www.dot.ca.gov/hq/oppd/proj_book/HCBP_2011a-9-13-11.pdf.

[20] Missouri DOT, "Core Team," *Engineering Policy Guide*.
Available at: http://epg.modot.mo.gov/index.php?title=104.1_Core_Team.

- Massachusetts DOT's (MassDOT) youMove Massachusetts is a transportation planning initiative under which workshops where held in 2009 throughout the state to gather over 700 public comments on mobility gaps. From these comments, MassDOT formulated themes and began identifying solutions to mobility challenges. One of the themes developed was directly related to designing for operations: "Theme 3: Design Transportation Systems Better – Transportation facilities and operations should be better informed by real-world conditions faced by system users." YouMove Massachusetts resulted in MassDOT adopting seven major components to their business practices, including "Transportation Reform," defined as: "Emphasis on our customers, innovation, accountability, performance management, efficiency, stewardship and stronger collaboration across transportation divisions."[21]

2.2 AGENCY STRUCTURE

In order to have a fully integrated view of operations, agencies are moving beyond a narrow construction and project viewpoint to one in which all of the customers' mobility needs are fully considered. A successful organizational approach to management and operations requires not just recognizing operations, but ensuring that operational strategies are formally considered during project development and infrastructure design. Key areas of an agency's structure that will contribute to an integrated program include agency culture and agency organization and staffing.

Illinois State Toll Highway Authority

Toll facility operators are traditionally customer-focused and uniquely concerned with operational needs during the design stages for successful tolling operations. The Illinois Tollway is organized around that philosophy. Its division of maintenance and traffic reports to the Chief Engineer, as does its design group. This helps to institutionalize more effective communications among operators and designers early in the design phase. This approach was further solidified when the engineering department became ISO-certified and the maintenance and traffic division became the "customer" of the engineering, design, and construction groups, ensuring operational input and approval by maintenance as part of the planning, design, and construction processes.

Source: *John L. Benda, General Manager of Maintenance & Traffic, Illinois Tollway.*

2.2.1 Culture and Leadership

Designing for operations requires a champion to voice the importance of operations within an agency. Key leadership must emphasize the importance of operations to the agency's customers. As State DOTs and other transportation agencies embrace their transition from the traditional functions of designing, constructing, and maintaining infrastructure to providing equal consideration of existing infrastructure, they will develop a closer relationship with the users of these systems. A customer-driven approach will further move the culture of the agency toward an operations focus. This shift in agency culture can provide an environment for designing for operations strategies to flourish.

2.2.2 Organization and Staffing

Agencies that are seeking to incorporate the practice of designing for operations into everyday activities will benefit from operations champions among top-level senior staff. Including operations into a formalized institutional structure will provide support to programs and projects in which operations is a core mission. In addition, raising the level of operations expertise through experience and staff training, including top agency management, will further elevate the operations functions within an organization.

A number of agencies have elevated operations within their organizations, which has provided better visibility for operations initiatives. For agencies interested in placing a greater emphasis on operations, a standalone operations division should be considered. Several agencies, including Virginia DOT and Minnesota DOT, have created dedicated operations divisions that are on the same organizational level as engineering/design within the organizational structure. Both of these agencies are at the forefront of incorporating operations into business practices, including design. The Florida DOT offers an alternative model by creating a TSM&O program that is integrated across the DOT's departments.

Several agencies combine operations with maintenance into a single division. While this provides an improved opportunity for collaboration, it may not provide top management with the highest level of visibility into the operations function.

---

[21] MassDOT, Transportation Planning Process. Available at: http://youmovemassachusetts.org/themes/design.html.

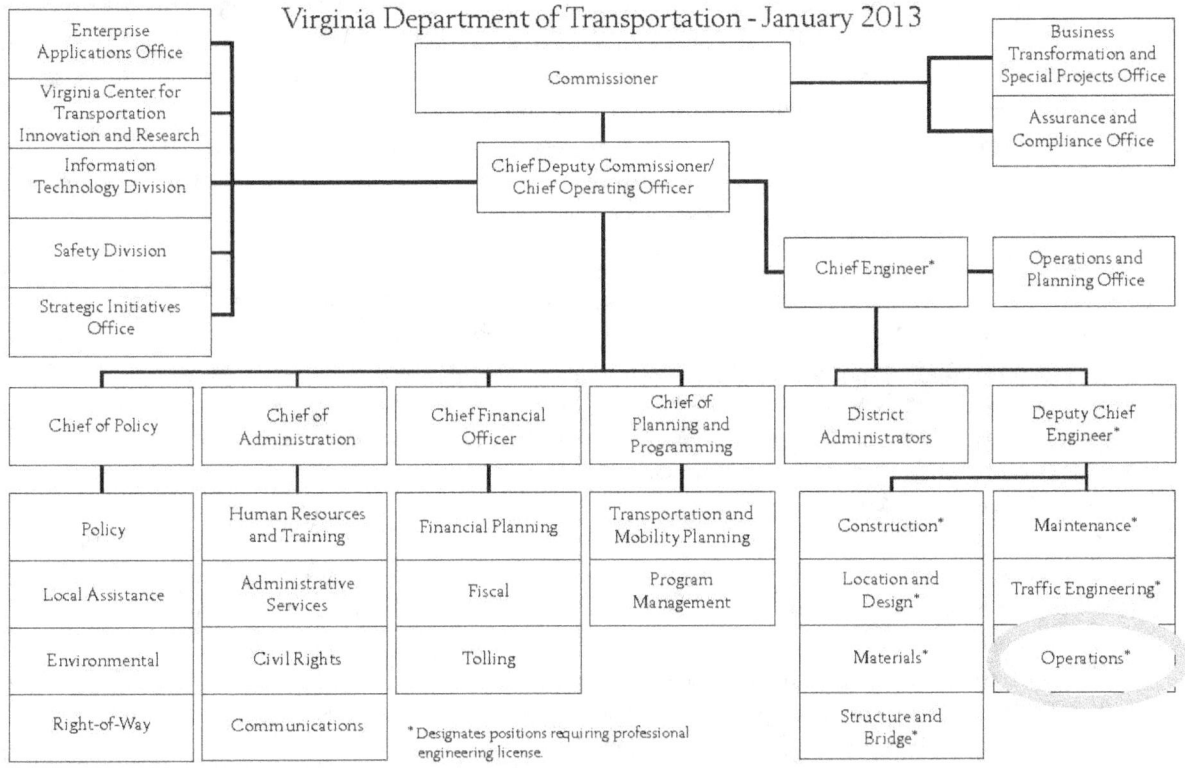

Figure 7. Virginia DOT organizational chart showing prominent operations division.[22]

2.3 LINKING PLANNING AND DESIGNING FOR OPERATIONS

An effective approach to mainstreaming the practice of designing for operations stems from a strong connection to planning at the State and metropolitan levels. The systematic inclusion of operations in transportation planning at the State and metropolitan levels provides a foundation for considering operations in project design. During the integration of operations into planning, known as "planning for operations," regional or statewide objectives and performance measures for the operation of the transportation system are established. In overview, these operations objectives and performance measures provide designers and operators direction and a specific purpose when considering how to incorporate operations into the design of a transportation facility.

Planning for operations is a topic of ongoing outreach and education by the FHWA. Planning for operations is defined as a joint effort between planners and operators to integrate management and operations strategies into the planning process for the purpose of improving regional transportation system efficiency, reliability, and options.

During the transportation planning process, guidance is collaboratively established for achieving desired outcomes for the region or State's transportation system in terms of visions, goals, objectives, and performance targets. Current and future issues and needs related to transportation are identified through public outreach, data collection, and modeling. Potential and preferred transportation solutions are developed and included in a long-range plan. The statewide plan is frequently more policy-oriented (rather than specifying a set of programs or projects), whereas in the case of metropolitan planning, a cost-feasible plan is developed. Based on this plan, projects are selected and prioritized for funding in the transportation improvement program (TIP) or statewide transportation improvement program (STIP).

By using a systematic approach to planning for operations, States and metropolitan regions can ensure that operations is taken into account in each of the phases of transportation planning that are described in the previous paragraph. The

---

[22] Virginia DOT, Virginia DOT's Organization, 2013. Available at: http://www.virginiadot.org/about/vdot_organization.asp.

systematic approach to planning for operations recommended by the FHWA is driven by specific, outcome-oriented objectives for the operational performance of the transportation system and is based on performance measures. **By linking an objectives-driven, performance-based approach to planning for operations to the project design process, infrastructure design can better reflect the needs, priorities, and performance targets developed through a collaborative process with operators, planners, and other stakeholders and agreed to by transportation decision-makers.** This means that highways, bridges, arterials, and rails are built to better meet the needs of the traveling public and businesses and are focused on delivering cost-effective performance.

Developing measureable, specific operations objectives is a significant part of the planning for operations approach that will have impacts on how infrastructure design incorporates operations. Each objective will typically include a performance measure and target to track its achievement. Operations objectives should have "SMART" characteristics: specific, measurable, agreed, realistic, and time-bound. Operations objectives should be selected that are reasonably achievable given limitations on funding and other demands. Defining operations objectives and performance measures is an important element of the planning process for designer participation. For example, if a region has an operations objective of "Attain a maximum variability of travel time on specified routes of XX percent during peak and off-peak periods by year 2020," then infrastructure designs will need to facilitate efficient TIM, work zone operations, and other strategies that will help achieve the reliability target. Alternatively, if the region or State selects the operations objective of "Achieve a per capita single occupancy vehicle commute trip rate less than YY percent in 15 years," infrastructure designs that increase the ease of using transit, car-sharing, bicycling, and walking will be a high priority. In addition to selecting outcome-based operations objectives, it is helpful to adopt supporting operations objectives or standards that focus on those aspects of the system that are directly controlled by transportation professionals, such as accuracy, reliability, and availability of traveler information and supporting ITS infrastructure. By being part of the development of operations objectives, designers can help ensure that the operations objectives and related performance measures are feasible given design constraints and provide input on opportunities for improving system performance through infrastructure design. After objectives have been established, designers can help promote designs that reflect the chosen operations objectives. Operations objectives can be linked to design and used to support the development of the purpose and needs documentation for the environmental approval process.

The diagram in Figure 8 illustrates the objectives-driven, performance-based approach to planning for operations and the important role of operations objectives. In the approach, M&O strategies are identified, evaluated, and selected based on operations objectives (where do we want to go?) and operations needs (where and why are we falling short?). Once selected, these strategies may be stand-alone projects carried out on existing infrastructure, or they may be incorporated into infrastructure expansion or reconstruction projects. Operations objectives, needs, and strategies identified and selected in the planning process can be inputs to the scoping and preliminary studies performed as a first step in project design.

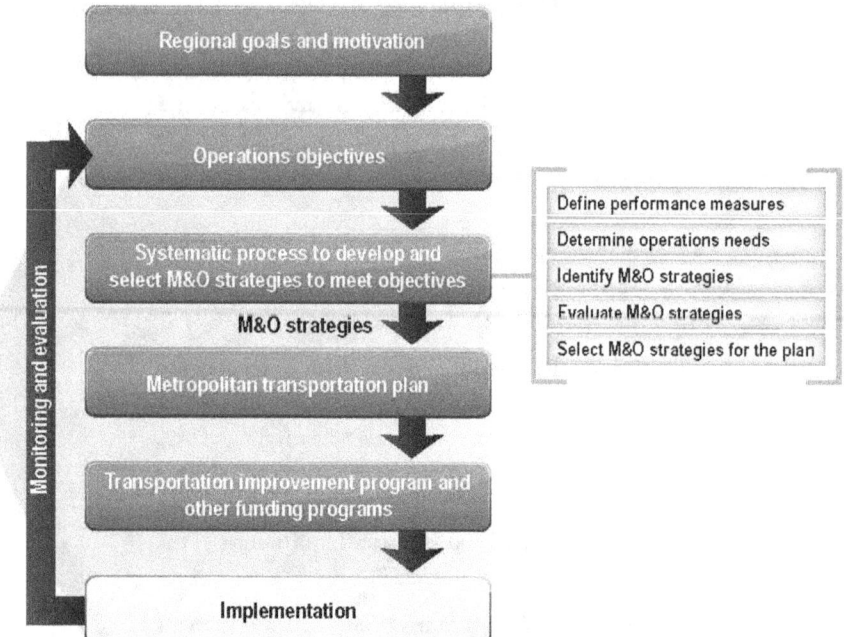

Figure 8. The objectives-driven, performance-based approach to planning for operations.

In addition to using the operations objectives, performance measures, and strategies from the transportation planning process, it is necessary for the regional ITS architecture to be part of including operations in the design process. A regional ITS architecture is a framework for institutional agreement and technical integration in a particular region. The architecture defines the links between the pieces of the ITS system and the data that is exchanged between systems. The U.S. DOT recently released a primer on opportunities to use the regional ITS architecture in planning for operations.[23]

It is necessary to examine the regional ITS architecture when developing the project scope in order to identify which ITS services exist or are planned for the region and how the technological component of the operations strategies considered for inclusion in the project can integrate with and benefit from those ITS deployments. This will help minimize the risk of disconnects in transportation services and will result in a project scope that considers the context of all other systems in the region. In addition to the benefits of using the regional ITS architecture in designing for operations, Federal regulation for Intelligent Transportation System Architecture and Standards Section 940 stipulates that all projects that fund the acquisition of technologies to provide an ITS user service (similar if not identical to operations strategy) with the highway trust fund shall conform to the National ITS Architecture and standards. The "final design of ITS projects funded with highway trust funds shall accommodate the interface requirements and information exchanges as specified in the regional ITS architecture."[24] Otherwise, the regional ITS architecture must be updated.

Other plans within a region or State should be reviewed during the project development process as well because they may contain operations strategies and other information critical to accounting for operational needs. These plans may include ITS strategic plans, regional concepts for transportation operations, freight plans, and safety plans.

2.4 PROJECT DEVELOPMENT PROCESS

By embracing the designing for operations approach to project delivery, enhancements to how operations are considered in the design process can be implemented throughout the various stages of design. The typical project design process includes scoping and financing, preliminary design, and final design stages, as shown in Figure 9. Each stage provides the opportunity to enhance the end product as it relates to operability and ease of maintenance of the constructed facility.

Successfully promoting operational needs and objectives throughout the project development process requires that design practitioners consider other perspectives in the context of the project development process, such as the perspectives of those who will ultimately be responsible for mobility, safety, and future constructability of the roadway. One way to achieve this is to encourage designers, maintenance staff, safety professionals, emergency responders, and traffic management center (TMC) operators to foster an understanding of their respective needs and priorities pertaining to roadway operability. Many agencies have an engineer-in-training rotation program where young engineers are exposed to a variety of divisions within an organization, providing an opportunity to view a project from multiple vantage points. This experience allows a new generation of the workforce to gain insight into and understanding of how design and operations fit together.

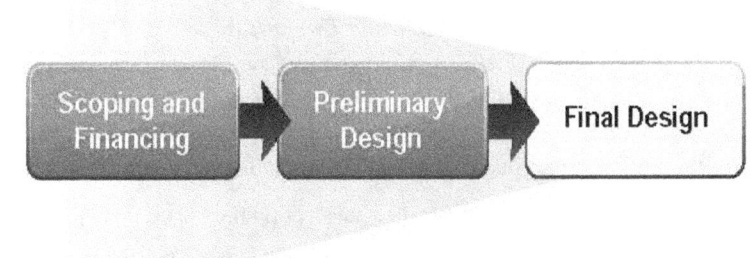

Figure 9. Generic steps in the project development process.

There are many coordinated project activities during the design phase, including environmental documentation, right-of-way acquisition, utility conflict assessment, structure and roadway geometric design, specification development,

---

[23] U.S. DOT, *Applying a Regional ITS Architecture to Support Planning for Operations: A Primer*, 2012, FHWA-HOP-12-001. Available at: http://www.ops.fhwa.dot.gov/publications/fhwahop12001/index.htm.

[24] FHWA and FTA Intelligent Transportation System Architecture and Standards. Available at: http://www.ops.fhwa.dot.gov/its_arch_imp/policy_1.htm.

final plan development, cost estimating, and project reviews. Soliciting and incorporating feedback from those familiar with operational concepts in the design and review processes may equip designers with an understanding of the needs for and benefits of management and operations on a typical project and when and where to incorporate the strategies.

2.4.1 Scoping & Financing Stage

During scoping and financing, practitioners will define project limits, establish a budget, and determine the project's schedule for subsequent design and construction phases. The project's fundamental purpose will also be identified, such as providing improvements to mobility for a corridor or subarea, addressing infrastructure repair or rehabilitation based on input from asset management data, enabling expansion of modal choice (e.g., bus, bicycle), or improvements to address a public safety concern.[25]

The opportunity to include operational considerations is greatest during the initial stages of a project. Agency operations and maintenance personnel and other partners (e.g., commercial vehicle operators, those tasked with emergency and incident response) may provide input on the scoping of a project. This stage presents a prime opportunity for practitioners to examine the need for additional ITS elements along a corridor. Project scoping staff can engage and collaborate with operations stakeholders by forming a design/operations/ITS committee that identifies specific ITS infrastructure needs along a corridor that will serve long-term mobility and safety goals. For example, during this stage emergency responders could provide input on the inclusion of staging areas or transit operators could give recommendations on bus rapid transit benefits and current and future infrastructure needs. Outside stakeholders can be engaged individually or through regional transportation operations working groups.

The project delivery method is also a consideration during the scoping and financing stage. With the proliferation of design-build and other fast-paced alternative delivery methods, collaboration with operational partners and incorporation of their considerations may be overlooked due to the involvement of many public and private entities.

2.4.2 Preliminary Design Stage

Preliminary design is the stage in which general project location and design concepts are determined and design element alternatives are considered. Preliminary design can include a wide range of preliminary engineering and other activities and analyses, including the National Environmental Policy Act (NEPA) process, geotechnical investigations, utility engineering, traffic studies, revenue estimates, financial plans, and others.[26]

In the very first stages of preliminary design, practitioners and the public will collaborate to determine the potential environmental impact a project may have. As a part of the NEPA process, practitioners must examine and avoid potential impacts to the social and natural environment when considering approval of proposed transportation projects. In addition to evaluating the potential environmental effects, agencies must also take into account the transportation needs of the public in reaching a decision that is in the best overall public interest. In the project development process, the NEPA process is an approach to balanced transportation decision-making that takes into account the potential impacts on the human and natural environment and the public's need for safe and efficient transportation.[27]

During the preliminary design stage, tangible operational considerations can be incorporated into the design documents. This stage presents a critical opportunity to solicit input from stakeholders—such as operations and maintenance personnel, emergency responders, and other end users of the facility—related to mobility and safety goals. Early in the preliminary design stage, when an array of project alternatives is still being examined, there is an opportunity to consider lower cost, operations-oriented improvements as alternatives to traditional infrastructure projects.

---

[25] FHWA, *Integrating the HSM into the Highway Project Development Process*.
Available at: http://safety.fhwa.dot.gov/hsm/hsm_integration/sec3.cfm.

[26] U.S. DOT, *Shortening Project Delivery Toolkit: Clarifying the Scope of Preliminary Design*.
Available at: http://www.fhwa.dot.gov/everydaycounts/projects/toolkit/design.cfm.

[27] For more information, see the FHWA Environmental Review Toolkit Website, NEPA and Project Development.
Available at: http://environment.fhwa.dot.gov/projdev/index.asp.

During this stage, agencies should review their ITS architecture and operations strategic plans to determine the feasibility of including elements from those plans as a part of the project contract. This leads to increased collaboration and communication between designers and operations staff as the group's interpretations of the ITS plans develop into finite design elements. It is also critical to collaborate with local county and city agencies, transit operators, and motor carriers to ensure that each stakeholder's long-term goals reasonably align with the roadway design. This type of collaboration and input can also pave the way for cost-share projects between agencies.

2.4.3 Final Design Stage

The final design stage is denoted by the preparation of construction plans and detailed specifications for construction work to be performed. During this stage, final plans will be developed that include traffic control and construction staging plans, exact quantities of known construction elements, an estimate of construction costs, and geographical coordinates for construction improvements.

While a project design has matured and developed by this stage, adjustments to the design to accommodate operations may still be considered. For example, local transportation and emergency response agencies may help to define the need for and location of temporary pull-offs for stalled vehicles within a freeway work zone where shoulders are not present.

The design process and operations do not move at the same speed. Some projects take years to move from preliminary design to final design to construction, while the technology used for operations changes on a continuous basis. Final design provides the opportunity to revisit operational considerations. For example, there may be a need to update a technical specification for a specific technology.

2.4.4 Examples of Designing for Operations in Project Development

Agencies that define the role of and promote input from fields of expertise outside of the typical project development process can see a variety of benefits in the short and long term. These benefits range from cost savings, reduced construction duration, collaboration across projects and areas of expertise, increased safety of roadway users and workers, and reduced litigation claims.

By including agency staff with varied backgrounds in the project development process, a cradle-to-grave project perspective may be impressed upon all who participate. Considerations borne from discussions and collaboration among the project team may result in a true understanding of the interdependence among seemingly separate elements of a project, resulting in the same or similar considerations being applied on future projects.

California Department of Transportation

Caltrans' *Project Development Procedures Manual*[28] recommends participation and input from various fields of expertise on each project development team during the planning, design, and construction phases. Operations are represented by both traffic and maintenance staff.

Traffic Operations. A representative from the District traffic unit serves on the project development team to provide input on traffic-related issues. During project planning, the traffic unit provides capacity studies and operational analyses and develops safety and delay indices. Traffic representatives determine whether the project alternatives will function adequately if constructed. Questions to be answered by the traffic unit during planning include the following:

- Is there sufficient room for hardware such as sign structures, electrical facilities?
- Should traffic signals, storage, and striping be considered?
- Is a transportation management plan needed?
- Have the results of the field safety review been incorporated when appropriate?

---

[28] Caltrans, *Project Development Procedures Manual (PDPM)*, 1999-2012. Available at: http://www.dot.ca.gov/hq/oppd/pdpm/pdpmn.htm.

During the design phase, the traffic unit is requested to review the geometric layouts to ensure that elements such as signing requirements, stage construction, intersection operation, end of freeway plans, and temporary connection plans are adequate for the safety of the motorists and construction and maintenance workers. The traffic unit is provided with skeleton layouts and requested to prepare the traffic-related portions of the project plans. This normally consists of the following elements:

- Traffic signing and striping plans;
- Lane closures and lane requirement charts;
- Traffic electrical plans including location of current transportation management system elements and stage construction;
- Traffic contract items and quantities;
- Signing and striping for traffic handling plans;
- Transportation management plans (TMP); and
- Special considerations unique to the project such as railroad signing.

At Caltrans, the District traffic unit's involvement in project development does not end with the award of a construction project. At various times throughout the construction project, the unit is expected to review closure schedule change requests, proposed traffic control measures, and signing and safety elements to ensure that public safety and convenience are considered. Stage construction, detours, and temporary connections may require modification to the TMPs, and changes are made in cooperation with the District TMP coordinator. The traffic unit is consulted prior to making changes in the TMP.

Designing for Ease of Maintenance. A Caltrans District maintenance representative serves on the project development team to ensure that maintenance issues and safety design are considered. Preferably, the representative will be the field person most familiar with the project site.

During project planning, maintenance involvement includes reviewing and commenting on features such as the following:

- Drainage patterns (e.g., known areas of flooding, debris);
- Stability of slopes and roadbed (i.e., can the project be built and maintained economically?);
- Possible material sites;
- Concerns of the local residents;
- Potential erosion problems;
- Facilities within the right-of-way that would affect alternative designs;
- Wildlife considerations (e.g., problems such as deer crossings, endangered species);
- Traffic operational problems (e.g., unreported accidents); and
- Safety of maintaining the facility.

In the design phase, the maintenance unit also reviews the proposed geometric layouts, typical sections, and final plans. Maintenance may have input on design details like shoulder backing materials, drainage, erosion control, access to buildings, access for landscape facilities, access to encroachments for utility facilities, and access for maintenance of noise barriers and fences. Maintenance staff also participates in the preparation of maintenance agreements (setting maintenance control limits).

The maintenance unit field representatives have unique insights into local problems and maintenance and safety concerns, bringing perspectives that can be utilized in the project development process. As the last link in the process, the maintenance unit can help minimize future maintenance problems and potential lawsuits.

Woodrow Wilson Bridge[29]

The Woodrow Wilson Bridge Project in Washington, D.C. was undertaken to replace an aging bridge structure and improve traffic flow on the I-95 corridor by doubling the number of lanes over the Potomac River. The project was one of the largest public works projects in the mid-Atlantic region and was sponsored by four cooperating agencies: FHWA, Virginia DOT, Maryland State Highway Administration (SHA), and District DOT (DDOT).

The design and construction phases of this large project were enhanced by the formation of an operations team, which included DOTs and external partners. The team met once per month to review key design submissions for integration into the project's TMP, which helped forge relationships that ultimately improved operations. Operational elements that were incorporated into the project as a result of the team meetings included standardized incident and emergency management features such as standpipes, hazardous materials (HAZMAT) sheds, and staging areas for service patrol vehicles.

2.5 SYSTEMS ENGINEERING

Systems engineering is an organized approach to developing and implementing a system. The approach can be applied to any system development, including an operations strategy on a roadway network. Whether deploying a few closed-circuit television (CCTV) cameras, upgrading your traffic signal system, or implementing active traffic management on a corridor, systems engineering can be used.[30] It is crucial to use the systems engineering approach in designing ITS infrastructure so that the technology effectively supports the management and operation of the transportation system. A systems engineering analysis is required for all ITS projects using Federal funds per Title 23 CFR 940.11.[31] The systems engineering approach helps to ensure that the system or operations strategy is responsive to the needs of all stakeholders, such as the traveling public, transit operators, businesses, incident responders, TMC operators, and others. The approach provides a systematic method for ITS and operations project developers to design their systems to achieve the desired operations objectives.

The International Council on Systems Engineering (INCOSE) defines systems engineering as follows:

Systems Engineering is an interdisciplinary approach and means to enable the realization of successful systems. It focuses on defining customer needs and required functionality early in the development cycle, documenting requirements, then proceeding with design synthesis and system validation while considering the complete problem.[32]

Prior to the design of a roadway project, operations and ITS staff should consider using the systems engineering process to systematically define how the roadway should be operated to achieve the region's operations objectives. This includes consulting the regional ITS architecture, identifying operations needs, and building a concept of operations for managing and operating the roadway if one does not already exist. This will be important input for operations considerations during the scoping and preliminary design stages of the roadway project. As illustrated in Figure 10, the traditional project development process (i.e., design-bid-build as opposed to design-build or other approach) parallels the systems engineering process, represented with the winged "V" (or "Vee") model. Designers can use the outputs of the early steps in the systems engineering process to guide the inclusion of operations considerations into design.

Following the "V" process from left to right, the left wing shows the regional ITS architecture, feasibility studies, and concept exploration that support initial identification and scoping of an ITS or operations project. As one moves down the left side of the "V," system definition progresses from a general user view of the system to a detailed specification of the system design. A series of documented baselines are established, including a concept of operations that defines the user

---

[29] U.S. DOT, *Livability in Transportation Guidebook*, Appendix 14. Virginia/Maryland - Woodrow Wilson Bridge, 2010, FHWA-HEP-10-028. Available at: http://www.fhwa.dot.gov/livability/case_studies/guidebook/.

[30] U.S. DOT, *Applying a Regional ITS Architecture to Support Planning for Operations: A Primer*, 2012, FHWA-HOP-12-001. Available at: http://www.ops.fhwa.dot.gov/publications/fhwahop12001/index.htm.

[31] Visit http://ops.fhwa.dot.gov/int_its_deployment/sys_eng.htm for additional information and resources, including the Systems Engineering Handbook (http://ops.fhwa.dot.gov/publications/seitsguide/index.htm) and *Systems Engineering Guidebook* (http://www.fhwa.dot.gov/cadiv/segb/).

[32] International Council on Systems Engineering, "What is Systems Engineering?" Web site, June 2004. Available at: http://www.incose.org/practice/whatissystemseng.aspx.

needs, a set of system requirements, and high-level and detailed design. The hardware and software are procured or built at the bottom of the "V," and the components of the system are integrated and verified on the right side. Ultimately, the completed system is validated to measure how well it meets the user's needs. The right wing includes the operations and maintenance, changes and upgrades, and ultimate retirement of the system.[33]

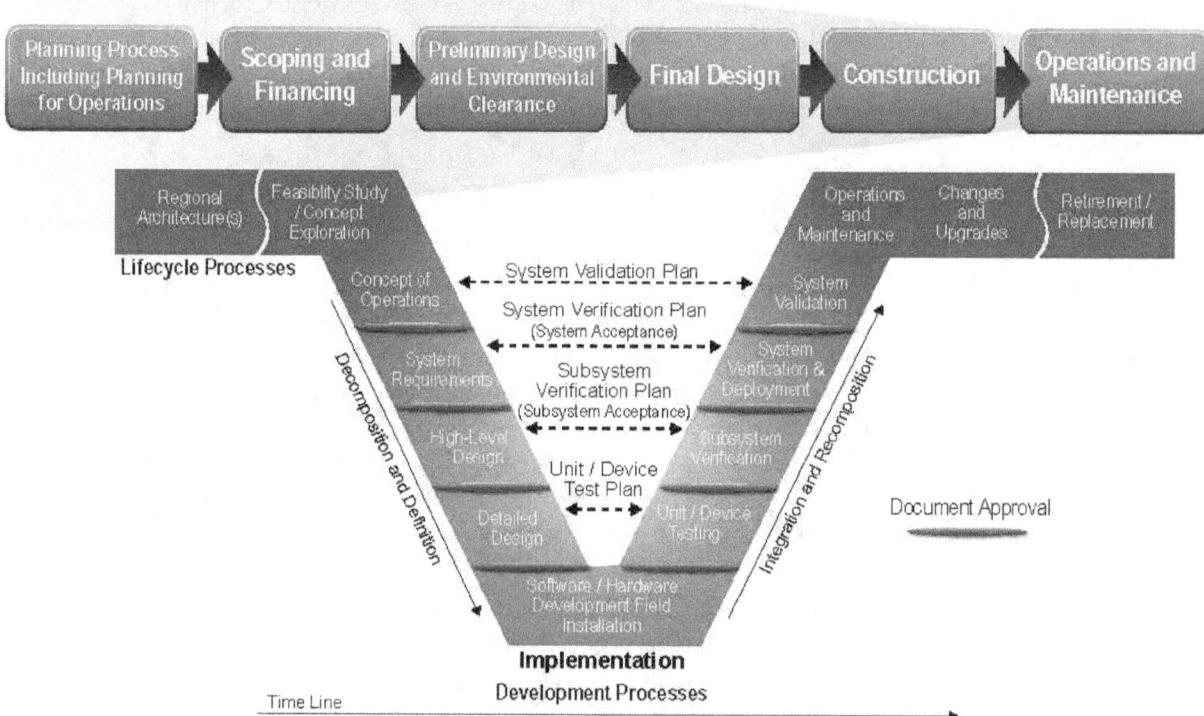

Figure 10. The systems engineering "V" model and traditional project development process.

2.6 DESIGN STANDARDS AND CHECKLISTS

2.6.1 One Stop Manual for Engineering and Other Technical Information

Bundling engineering and technical information into one document can reduce barriers to incorporating operations into design. Designers are able to access basic information on operations policies, strategies, and treatments related to items such as access management standards, preventive maintenance policies, weather-related treatments and action plans, TIM, and operational policies and performance measures for the use and reliability of ITS elements.

> Washington State DOT has developed an operations design matrix used to evaluate the impact of operational elements. As a result, there are facilitated discussions regarding types of operational elements to be included in a project.

In addition, having a "one-stop" document or manual prevents the common error of creating duplicative policies with differing outcomes, performance expectations, and operational objectives. The document will provide the same interpretation of external resources used to develop agency policies, and the policies within it can be updated as new research, practices, and innovations become widely accepted. Lastly, the document will provide common language and terminology to be used throughout an agency's divisions, offices, or districts.

---

[33] U.S. DOT, *Applying a Regional ITS Architecture to Support Planning for Operations: A Primer*, 2012, FHWA-HOP-12-001. Available at: http://www.ops.fhwa.dot.gov/publications/fhwahop12001/index.htm.

Integrating operations into design could be supported by checklists for standalone operations projects and to encourage the incorporation of operational strategies into larger projects. Pennsylvania DOT has developed design checklists for a number of ITS elements for both standalone projects and components of a larger project (see Figure 12). The checklists are part of the Pennsylvania DOT Intelligent Transportation Design Guide and address the design of CCTVs, dynamic message signs, highway advisory radio, vehicle detectors, ramp meters, and travel time systems. The checklists ensure that a thorough list of location, safety, power, communications, maintenance, usability, and other factors or requirements have been considered in the design of the ITS element. The checklists help to ensure an effective, consistent, and cost-efficient application of ITS as well as a design that is consistent with operations needs and the regional ITS architecture(s).

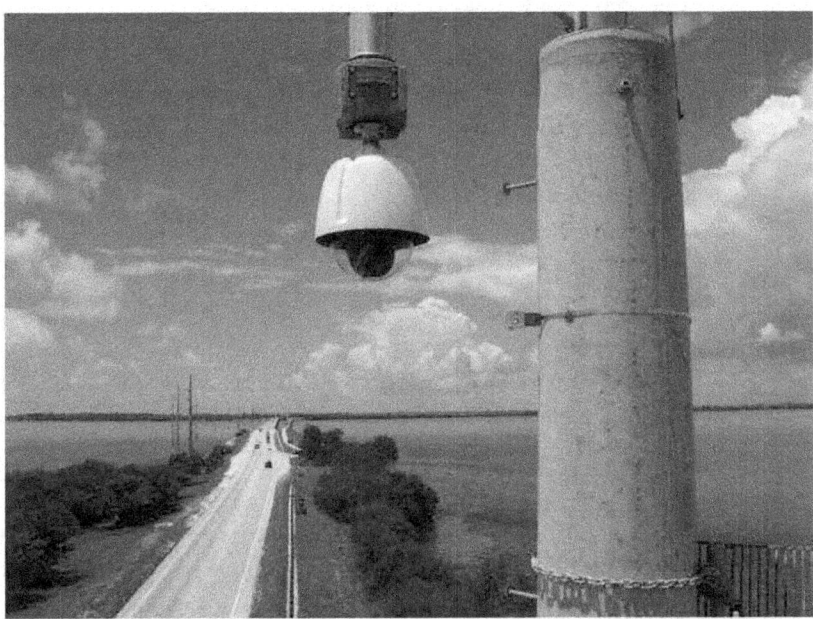

Figure 11. The use of checklists can help ensure the CCTV cameras and other ITS equipment effectively support operations.
(Source: Florida Department of Transportation)

pennsylvania — *PennDOT Publication 646 Intelligent Transportation Systems Design Guide*

10 Appendix A – Design Checklists

10.1 CCTV Design Checklist

Detection Purpose	YES	NO	N/A
• Is this deployment consistent with "needs" outlined in a Concept of Operations or Regional Operations Plan?	☐	☐	☐
• Is this deployment consistent with the ITS architecture?	☐	☐	☐
Location/Placement Guidelines	**YES**	**NO**	**N/A**
• Has the camera location been chosen/designed with consideration to maximizing visibility?	☐	☐	☐
• Has a site for the camera been chosen that considers the available utilities and the cost/constraints associated with connection to those utilities?	☐	☐	☐
• Has the site been chosen with consideration to protecting the camera structure and ensuring that it will last; without undue maintenance necessary to the structure and the surrounding site?	☐	☐	☐
• Has a site been chosen that makes the best use of the operational needs of a CCTV camera system (e.g. Incident Management)?	☐	☐	☐
• Has a site been chosen that satisfies safety requirements for personnel performing maintenance on the system?	☐	☐	☐
• Has the site been chosen so that it will minimize maintenance costs (e.g. there is sufficient shoulder to park a bucket truck without the need for a full lane closure and significant MPT)?	☐	☐	☐
CCTV Type	**YES**	**NO**	**N/A**
• Is the camera type (external vs. dome) appropriate for the desired location and application?	☐	☐	☐
• Does the District require the camera to be compatible with a legacy analog system?	☐	☐	☐
Camera Mount	**YES**	**NO**	**N/A**
• Have Department standards ITS-1000M and ITS-1002M been followed in the design of the mount/structure?	☐	☐	☐

pennsylvania — *PennDOT Publication 646 Intelligent Transportation Systems Design Guide*

Enclosure	YES	NO	N/A
• Is an enclosure required at this location?	☐	☐	☐
• Can personnel safely access the enclosure?	☐	☐	☐
• Is the enclosure located within 150 feet of the camera?	☐	☐	☐
• Is the enclosure mounted on the camera pole or on an existing structure (where possible)?	☐	☐	☐
• Does the location and orientation provide adequate protection for the enclosure?	☐	☐	☐
• Has a concrete maintainer's pad been provided at the enclosure's main door?	☐	☐	☐
Power Requirements	**YES**	**NO**	**N/A**
• Have the power requirements for the camera and all of the system components been determined?	☐	☐	☐
Power Availability	**YES**	**NO**	**N/A**
• Has an appropriate power source been located and confirmed with the utility company within a reasonable distance from the camera site?	☐	☐	☐
• Have Step-Up/Step-Down requirement calculations been performed?	☐	☐	☐
• Have the metering options been determined?	☐	☐	☐
Power Conditioning	**YES**	**NO**	**N/A**
• Do the standard grounding specifications meet the needs of the system?	☐	☐	☐
• Have the UPS and power back-up options been determined and accounted for?	☐	☐	☐
Communication	**YES**	**NO**	**N/A**
• Have the communication requirements for the camera been determined?	☐	☐	☐
• Has an appropriate communication infrastructure been located and confirmed within a reasonable proximity to the site?	☐	☐	☐
• If there are multiple communication options, have the pros/cons been studied?	☐	☐	☐
• If using public communications infrastructure, has service been coordinated with PennDOT BIO and the District?	☐	☐	☐

PUB 646 ITS Design Guide — Page 112

Figure 12. Pages from Pennsylvania DOT publication 646, *Intelligent Transportation Systems Design Guide*, showing CCTV design checklist.

2.6.2 Operational Review and Sign-off of Standard Plans and Specifications

Establishing a review process in which design, maintenance, operations, and construction staff provide input and comments on the design ensures that operational impacts will be accounted for in the development of standard plans and commonly used specifications. The table below highlights the key project development phases for traditional design projects and notes the key documents that should be reviewed as part of an "Operational Review Process." Documented needs that are not addressed should require some type of approval or sign-off by leadership that clearly explains why the operations strategy is not being included. This sign off could be similar to a "design exception."

The table below demonstrates how two different types of reviews can be applied to the traditional project development process to ensure operations considerations are being considered throughout the design process. Traceability reviews consider how a specific project is addressing high-level goals and strategies developed as part of a high-level plan or policy document such as a regional operations strategic plan or a long range transportation plan. These high-level documents may make overarching statements like "incorporate advanced technologies on new projects to improve traveler information." The traceability review would serve as the checkpoint to ensure that these high-level statements are being considered for specific projects.

The operational reviews are specific to a particular project. The recommendation is that for projects of significance, the project team would develop a technical memorandum that identifies specific operations strategies. The definition of "significance" can be established by each individual agency.

Table 1. Example reviews for project development to ensure consideration of operations.

Project Development Phase	Traceability Reviews	Operational Reviews
Project Scoping	• Mobility plans. • TIM strategic plans. • Long range transportation plans and associated operations objectives, performance measures, and M&O strategies. • Regional operations plans and/or regional ITS architectures. • Congestion management plans. • Existing corridor operations plans.	• Develop project-specific operational strategies memorandum. • Obtain concurrence by operational review team. • Approval and sign-off by agency leadership.
Preliminary Engineering and Environmental Documentation	• Review operational strategies memorandum created during scoping phase.	• Concurrence by operational review team. • Approval and sign-off by agency leadership.
Final Design	• Verify "needs" documents identified above have been addressed. • Review project plans, specifications and system engineering documents.	• Concurrence by operational review team. • Approval and sign-off by agency leadership.

The **Strategic Highway Research Program 2 (SHRP 2)** is currently concluding a 4-year project on the "**Evaluation of the Costs and Effectiveness of Highway Design Features to Improve Travel Time Reliability.**" The results of this effort (SHRP 2-L07) are expected to greatly improve the capability of a transportation agency to design for operations. The project will produce a guidebook and analysis tool that will help predict the operational and safety benefits of a range of design treatments. The guidebook is anticipated to include descriptions and examples of highway design treatments that reduce non-recurring congestion, quantitative and qualitative safety and operational benefits of the treatments, and assistance on how to evaluate and select design treatments.

2.6.3 Operations Audits and Review Team

As stated previously, communication among planners, designers, and operations professionals is of key crucial importance to the successful integration of operations into design. The utilization of operational reviews and audits during design can help ensure that operational strategies are being considered and that the project ultimately meets the operational objectives identified during planning.

A successful model of this concept already exists. Some States have incorporated safety reviews during key project milestones. The inclusion of operations would be similar and would include the input of planners, designers, and operations professionals during project scoping, preliminary engineering, and final design. The operational review team should be an independent team if possible to ensure a review is conducted objectively.

3 Design Considerations for Specific Types of Operations Strategies

Proper design of operational elements and inclusion of M&O considerations in typical infrastructure projects provide an opportunity to maximize the efficiency of a transportation system. Future deployments can be jeopardized when operations considerations and provisions are not included in projects. This chapter provides an overview of M&O strategies and considerations for incorporating operations into the design of transportation projects. It is intended to help designers understand what design issues may be associated with specific operations strategies and how to include these strategies on typical transportation infrastructure projects. It should be noted that many of these strategies can also be deployed on a standalone basis where appropriate. Many strategies and design considerations described in this chapter are repeated in various sections due to the overlap and interdependence of these strategies. For example, freeway management and arterial management are essentially broader categories that include elements of subsequent sections of the chapter, such as managed lanes, active traffic management, and maintenance. The decision of which operations strategies to consider during the design of projects may often be driven by overarching operations objectives and a concept or plan for managing and operating the transportation system.

3.1 FREEWAY MANAGEMENT

Freeway operations and traffic management involve managing travel and controlling traffic. The application of appropriate policies, strategies, and actions can mitigate any potential impacts resulting from the intensity, timing, and location of travel and can enhance mobility on highway and freeway facilities. Freeway management systems can improve the efficiency of available capacity, improve safety, and support TIM activities. These systems can also be used to mitigate existing features in the cases of curve warning systems or runaway truck ramps.

Figure 13. Runaway truck ramp on Interstate 24 in Tennessee, a design element that supports highway operations. (Source: Tennessee Department of Transportation)

The FHWA's *Freeway Management and Operations Handbook* states: "Freeway traffic management and operations is the implementation of policies, strategies and technologies to improve freeway performance. The over-riding objectives of freeway management programs are to minimize congestion (and its side effects), improve safety, enhance overall mobility, and provide support to other agencies during emergencies." By following the strategies and design considerations described in this chapter, designers can support the strategies described in the Handbook.[34]

During the design phase for freeways, many operational strategies need to be accommodated. Design for non-recurring congestion caused by weather events, accidents, construction, emergency repairs, and other events must be integral to the physical design of the facility.

---

[34] FHWA, *Freeway Management and Operations Handbook*, Section 1.1 "Scope of Freeway Management and Operations," June 2006. Available at: http://ops.fhwa.dot.gov/freewaymgmt/publications/frwy_mgmt_handbook/chapter1_01.htm#1-1.

As technology has become more integrated with the transportation system, the opportunity to utilize ITS and other means to manage freeways has become more prevalent. Various devices and systems placed on freeways, including CCTV, dynamic message signs (DMS), and ramp meters, have changed the way freeways are operated, but have not adequately changed the way they are designed. ITS and other technologies are too often treated as an afterthought in the design process; designers tend to "fit them in" rather than design optimal locations for them.

The challenge now is not just to include technology in a project, but make it a seamlessly integrated portion of the design, similar to the design of stormwater management, utilities, or guiderail. When ITS is included on a new or reconstructed facility, efforts should be made to integrate the devices and communications into the overall design of the facility in order to ensure optimal placement of the devices. Since ITS is used to monitor and manage the freeway, the locations of devices are crucial. For example, a DMS displaying a message to travelers about the congestion they are currently sitting in is not located in a place where it can have maximum effectiveness. Rather than locating the sign where there is available right-of-way, the sign should be located to improve the operation of the facility. Considerations such as sun glare, guide sign spacing, spacing from the next interchange, and visibility due to horizontal or vertical curvature are just a few of the design considerations for placement of a DMS, as well as for other freeway management strategies. To embrace designing for operations, designers must explore these considerations within the framework of the overall design of the facility.

Table 1 identifies elements to consider during design that can impact freeway operations. It also shows potential opportunities for designers to structure their roadway design (or redesign) to allow for more cost-effective implementation of freeway management strategies in the future. Some of these design considerations would apply to multiple strategies.

An additional resource to supplement the design considerations for ramp meters listed in the table below is the FHWA *Ramp Management and Control Handbook*.[35] This reference contains a section focused on design considerations for ramp metering based on a variety of ramp metering design manuals and guides from across the United States.

Figure 14.
One-tenth mile markers support incident management on freeways.
(Source: MUTCD)

Table 2. Example design considerations and opportunities for various freeway management strategies.

Freeway Management Strategy	Design Considerations/Opportunities
Strategic Highway Safety Plans or Toward Zero Deaths Efforts	• Consult operations staff about rumble stripes/strips due to noise, pavement age and thickness, and marking. • Wider rumble stripes require retrofitting existing equipment.
Managing Non-recurring Congestion	• Include signing for routing incident-related traffic through adjacent arterials. • Include emergency refuge or pull-off areas and crash investigation sites. • Provide for large-scale evacuation through contra flow lanes and appropriate signing. • Include detection to activate special signal timing schemes on adjacent arterials for traffic diverted off the freeway. • Provide median breaks, crash investigation sites, and permanent crossovers at major bridges. • Provide 1/10 or 2/10 mile markers and other structure identifiers for the motorist to support incident detection and TIM.
Ramp Meters	• Consult with arterial road operators to determine the best way to avoid queues on the feeding arterials. Allow for adequate width in the original design to accommodate future HOV bypass lanes.
Traveler Information	• Incorporate information related to transit operations, such as park-and-ride lot locations prior to bottleneck locations. • Provide travel time information for all available modes of transportation, including light rail, bus, and subway. • Build areas to allow portable changeable message signs (CMS) to be deployed due to DMS outages and repairs.

---

[35] FHWA, *Ramp Management and Control Handbook*, January 2006, FHWA-HOP-06-001. Available at: http://ops.fhwa.dot.gov/publications/ramp_mgmt_handbook/manual/manual/index.htm.

Freeway Management Strategy	Design Considerations/Opportunities
Managing Weather Events	• Consider anti-icing devices for bridges. • Consider locations for full road weather information system sites or individual components for specific conditions (e.g., wind or fog). • Consider providing storage sites for maintenance.
ITS and Communications Technology	• Install conduit for fiber optic networks and expansion of communications devices for ITS and other technology along the freeway. • Provide adequate points of access for ITS devices as well as other agencies' needs, such as automated enforcement.

Figure 15. Road Weather Information Systems (RWIS) may be considered to support freeway management during weather events. (Source: Tennessee Department of Transportation)

Figure 16. Gates to manage highways during weather events or emergencies may be considered during design. (Source: Tennessee Department of Transportation)

3.2 ARTERIAL MANAGEMENT

Arterial management involves implementing practices and operations strategies that promote the safe and efficient use of arterial roadway capacity to manage congestion. It also promotes the idea of treating the transportation system as a network that serves transit, bicycles, and pedestrians in addition to motorists. Improved modeling capabilities have improved understanding of how the transportation system is a connected network: what happens in one location affects another. Design of freeway, arterial, and bridge projects must consider impacts on the operations of the local transportation network. Agencies must work together regardless of jurisdiction to ensure the proper strategies are put in place to mitigate the impacts on the surrounding network.

Successfully managing the safety and performance of arterials involves the following core functions:

- Cooperation of municipalities;
- Managing access for all modes; and
- Monitoring and actively managing traffic conditions and intersection signalization.

Other arterial management strategies to consider include traffic management during construction (alternate/detour routes), turn lanes, bus turnouts, crosswalk layout, and alternative intersection designs, such as displaced left-turn intersections and U-turn intersections.

Caltrans applies ramp metering as a crucial operational strategy for managing traffic and has developed two documents to guide project developers and designers in the planning and design of ramp meters. The Caltrans **Ramp Metering Design Manual** defines the "traffic operational policies, design standards and practices for ramp metering installations at new or existing entrance ramps," whereas the second document, the **Ramp Metering Development Plan**, establishes a list of each ramp meter currently in operation or planned over the next 10 years throughout California. The development plan is framed as a tool to facilitate coordination between functional units in Caltrans and with external partners in the planning and programming of ramp meters. Caltrans has also incorporated ramp metering into its statewide training courses to help integrate ramp metering throughout the project planning, design, and construction process. The ramp metering design manual contains information on storage length, HOV preferential lane, modifications to existing HOV preferential lanes, enforcement areas, and maintenance pullouts.

The ramp metering design manual instructs project development teams to consult the District Operations Branch before beginning any ramp meter design thus encouraging cross-functional collaboration. The manual also indicates that any freeway segment identified within the development plan should include provisions for ramp meters.

For more information, see: *http://www.dot.ca.gov/hq/traffops/systemops/ramp_meter/.*

3.2.1 Cooperation of Municipalities

Because arterials often fall under the jurisdiction of different agencies, managing arterials properly requires cooperation and collaboration with neighboring communities. A project under one jurisdiction should achieve a level of operation similar to the rest of the corridor. In order to accomplish this, agencies may need to form agreements. For example, to move traffic through the signals on a multi-jurisdictional corridor in order to maintain traffic flow, a designer may need to connect to another agency's network to share intersection data, share information on preemption for transit and emergency vehicles, or consider special event timing plans.

These agreements can be formal concepts of operations or memoranda of understanding, or they can be informal "hand-shake" agreements that have been institutionalized through years of effort. Designers must educate themselves on the content of these documents or other collaborative arrangements to understand how a project may impact the arterial as a whole. These agreements help agencies share a common language regarding operational goals, performance measures, and strategies to manage the arterial.

3.2.2 Managing Access for All Modes

Managing access is a primary strategy for improving operations on an arterial. Many agencies have guidelines on design elements such as driveway spacing, corner clearance from major intersections or interchanges, and the optimum location of signals and roundabouts. While these provide foundational knowledge to apply to arterials, designers cannot expect to follow the guidelines exactly, as adjustments are often necessary.

Figure 17. Example of a facility design that did not account for pedestrians or bicycles. (Source: SAIC)

Intersection Control Evaluation (ICE) is a process by which the most appropriate traffic control is selected through a holistic decisionmaking framework. Significant intersections are targeted, and impacted agencies provide feedback on which traffic control strategies to deploy (e.g., traffic signal vs. roundabout). This process helps support context sensitive solutions such as road diets and complete streets. Minnesota DOT practices ICE and has an Intersection Control Evaluation Guidelines for Implementation document. Caltrans is considering adopting the practice as well.

Source: http://www.dot.state.mn.us/trafficeng/safety/ice/index.html

Designers should check for operational impacts due to deviations from these guidelines. For example, when implementing context-sensitive solutions such as road diets and adding bicycle facilities, the operational impacts of the following should be considered:

- Intersection traffic control (e.g., pre-timed, actuated coordinated, closed loop, adaptive control, roundabouts)— Selection requires a detailed analysis to balance cost, travel time, and delay for all modes as well as other defined operational parameters.
- Median treatments (e.g., pedestrian refuge, center turn lanes, raised medians)— These treatments can impact the overall safety for all users of the facility, access to adjoining property, and efficiency for all modes.
- Multimodal transportation facilities— this including bus stops or turnouts and bicycle lanes.

3.2.3 Monitoring and Actively Managing Traffic Conditions and Intersection Signalization

Designers planning physical changes to an arterial roadway will need expertise in traffic operations in order to evaluate these changes because a simple report about level of service will not adequately address operational issues. Proper design and management of intersection traffic signalization is essential to optimizing the operation of arterial roadways. Designers must consider the overall corridor and roadway network signalization concept when designing a project for individual intersections. Questions about managing queues, operating speeds, safety modeling, and impacts due to growth in traffic and increases in pedestrian and bicycle modes need to be addressed. Designers should provide for in-pavement loops or other traffic monitoring devices to allow for operational assessments, including signal timing and progression.

Table 3 identifies elements to consider during design that can impact arterial operations. It also shows potential opportunities for designers to structure their roadway design (or redesign) to allow for more cost-effective implementation of arterial management strategies in the future. Some of these design considerations would apply to multiple strategies.

Figure 18. The photo highlights the importance of considering safe, efficient pedestrian access to bus stops as part of a complete arterial management strategy. (Source: SAIC)

Table 3. Example design considerations and opportunities for various arterial management strategies.

Arterial Management Strategy	Design Considerations/Opportunities
Strategic Highway Safety Plans or Toward Zero Deaths Efforts	• Type of median treatment for passing lane configurations. • Use of rumble stripes/strips.
Collaboration of Agencies and Municipalities	• Designers can facilitate regional operational practices and procedures by providing technical information to support multi-agency agreements. For example, designers can provide intersection dimensions for a system-wide change of clearance intervals of signalized intersections. • Support existing maintenance agreements between jurisdictions through infrastructure design related to snow removal, striping, signal maintenance and repair, roadway surface repair, permitting, and drainage as networks cross jurisdictional boundaries. • Seek out informal but institutional arrangements related to management and operation of the corridor and advance them into standards or executed agreements. • Uphold the principles and performance measures established in any concept of operations being used to govern the management of the corridor.
Manage Access	• Use traffic modeling to assess changes in access management near signals and other major intersections. • Consult expertise in traffic operations to evaluate the impacts of adjusting access due to actual site conditions. • Have designers and operators jointly review redevelopment proposals containing changes in access to be sure transportation needs are met (e.g., road diets).
Intersection Control	• Establish operations objectives and performance measures related to queue management, storage requirements, multi-modal impacts, and turning restrictions. • Use the FHWA systems engineering process to evaluate the appropriate signal system for progression (e.g., actuation, closed loop, or adaptive signal control). • Consult with operations staff about the location of signal hardware for ease and safety of maintenance.
Signal Coordination, Traffic Responsive Intersection Control, and Adaptive Signal Control Technology (ASCT)	• Provide traffic monitoring devices to allow for optimum operations and signal timing and progression. • For regional traffic signal systems, designers must consider how communications and maintenance will be managed since multiple agencies may be responsible for a single system. Agreements between agencies should be developed during the design stage to address these issues.
Context Sensitive Solutions (e.g., complete streets)	• When constructing or upgrading sidewalks, eliminate other barriers to pedestrian access by adding countdown pedestrian signals, pedestrian ramps, and associated hardware and conduit for these treatments. • Contact the appropriate department or agency to update pedestrian timing at signals. • Facilitate transit operations by implementing strategies such as bus turnouts, preemption for buses, and directional signing of transit facilities.

3.3 ACTIVE TRAFFIC MANAGEMENT

Active traffic management (ATM) and managed lanes (see Section 3.4) are becoming increasingly popular in the United States as facility operators seek innovative solutions that can improve throughput and safety on congested facilities largely within the footprint of existing highways, thus requiring little or no roadway widening.

ATM is the dynamic management of recurrent and non-recurrent congestion based on current and forecasted traffic conditions. ATM focuses on maximizing trip reliability through approaches that seek to increase throughput and safety through the use of integrated systems and new technology. ATM includes the automatic and dynamic deployment of M&O strategies to optimize performance quickly and avoid the delay that occurs with manual deployment.

Some ATM strategies, such as ramp metering and variable speed limits, have been successfully implemented within many parts of the United States. Most other ATM strategies are relatively new concepts in the United States; however, they have been successfully implemented in many parts of Europe.

Figure 19. ATM lane control signage on a highway in Washington is placed in a high-visibility location to optimize operations. (Source: Washington State DOT)

Figure 20. Variable speed limit sign in fog warning zone on I-75 near Cleveland, TN. Speed limit is reduced during periods of fog. (Source: Tennessee Department of Transportation)

ATM strategies fall under the broader context of active transportation and demand management (ATDM). ATDM is the dynamic management, control, and influence of travel demand, traffic demand, and traffic flow on transportation facilities. Available tools and assets are used to manage traffic flow and influence traveler behavior in real-time to achieve operational objectives, such as preventing or delaying breakdown conditions, improving safety, promoting sustainable travel modes, reducing emissions, or maximizing system efficiency.

Designers must be prepared and trained for this new environment in transportation as ATM installations may require changes to the geometry of the roadway. This may include the design of gantries to support both static sign and changeable lane use control signs for ATM applications. Table 4 identifies elements to consider during design that can impact ATM operations. It also shows potential opportunities for designers to structure their roadway design (or redesign) to allow for more cost-effective implementation of ATM strategies in the future. Some of these design considerations apply to multiple strategies.

Table 4. Example design considerations and opportunities for various ATM strategies.

ATM Strategy	Design Considerations/Opportunities
Dynamic Speed Limits (DSL)	• During original roadway and ITS design, provide adequate conduit in the median barrier or shoulder to accommodate future DSL signage. • Consider line-of-sight impacts in placement of DSL signs. • Consider how DSL signs will compete with other signs.
Speed Harmonization	• If gantries are used, locate periodic overhead signage that takes into account how sight distance is affected by vertical/horizontal alignment, the ease/expense of retrofitting with sign foundations, and required spacing for messaging. • Consider catwalks or other means of maintaining equipment while limiting lane closures.
Dynamic Lane Use Control	• During placement, consider special geometric characteristics and driver decision points. • Consider line-of-sight impacts in placement of lane control signs. • Ensure that lane control symbol options (text, symbols) comply with the Manual on Uniform Traffic Control Devices (MUTCD). • Consider catwalks or other means of maintaining equipment while limiting lane closures.
Dynamic Shoulder Lanes (Hard Shoulder Running)	• Provide emergency pull-off areas where right-of-way allows. • Design exceptions for geometric standards, including lane width, vertical and lateral clearance, and stopping sight distance may be required. • Consider drainage structures and storm water/snow storage, including inlet grates (motorcycle safety). • Striping of shoulder lanes must comply with the MUTCD (outside edge and separation between general purpose and shoulder lane). • In deciding whether to utilize left- or right-side shoulder, analyze primary access points, especially for bus-on-shoulder lanes. • Consider site-specific criteria when designing for safe crossing of ramps at interchanges. • Account for speed differentials between dynamic shoulder lane and general purpose lane. • Provide CCTV coverage to make sure lanes are clear of vehicles and debris. • Consider providing additional static signing.
Queue Warning	• Locate signage in advance of locations where queues typically form.
Traffic Surveillance and Incident Management	• Design and construct CCTV in high-crash locations to improve detection and verification time. • Provide maintenance access to CCTVs.
Adaptive Ramp Metering	• Allow for adequate width in the original design to accommodate future HOV bypass lanes. • Provide maximum available approach lane for vehicle storage to avoid backing up onto intersecting arterials.
Dynamic Junction Control	• Requires traffic information to operate the strategy. Data regarding maximum capacity of upstream lanes; traffic volumes on highway lanes and merging ramps; travel speeds on highway lanes and merging ramps; and incident presence and location are essential. • Optimally, include an expert system to deploy the strategy based on prevailing roadway conditions without requiring operator intervention. • Dynamic merge control requires overhead electronic signage.

3.4 MANAGED LANES

Managed lanes are highway facilities or a set of lanes where operational strategies are proactively implemented and managed in response to changing conditions. Managed lane projects take lane management strategies that have been used extensively for decades—such as HOV lanes, bus-only lanes, truck lane restrictions, and express lanes—and incorporate the concept of active management. These strategies can be implemented utilizing concurrent flow lanes (adjacent to general purpose lanes), reversible flow lanes, contra flow lanes, or existing shoulders. Colorado DOT took enforcement needs into consideration when designing their high-occupancy toll (HOT) lanes by adding a widened shoulder (see Figure 21).

Figure 21. Colorado I-25 HOT lanes enforcement shoulder. (Source: Myron Swisher)

Impacts of not considering operations during design. It is common for ITS specifications to provide recommended spacing of certain devices, and designers have effectively applied such spacing on traditional highway projects. Managed Lane and ATM applications require much more precise locations to fit project-specific needs. For example, generic placement of variable message signage could lead to providing real-time traffic and toll rate information beyond the point on the facility where a driver can use the information to make a route choice decision. Furthermore, the proliferation of information available through ATM applications can lead to confusion and apathy on the part of the driver. The information must be delivered in a very specific sequence and location in order to be of most use to the driver and operating agency.

Caltrans issued guidance in 2011 titled "Traffic Operations Policy Directive 11-02 – Managed Lane Design," which institutionalizes the practice of designing for operations. The directive states that it "shall be applied during the planning and development of freeway managed lane projects, including conversion of existing managed lanes to incorporate tolling or utilize continuous access. It shall be considered during the planning and development of all other freeway improvement projects (e.g., pavement rehabilitation project) and during the course of traffic investigations that are addressing operational and safety performance deficiencies."[36]

Table 5 identifies elements to consider during design that can impact managed lane operations. It also shows potential opportunities for designers to structure their roadway design (or redesign) to allow for more cost-effective implementation of managed lane strategies in the future. Some of these design considerations would apply to multiple strategies.

Table 5. Example design considerations and opportunities for various managed lane strategies.

Managed Lane Strategy	Design Considerations/Opportunities
HOT Lanes	• In order to optimize transit use of a managed lane facility, consider major bus routes when locating weave zones to enter and exit the lanes. • Access/egress zones for buffer separated facilities must be carefully located, with consideration given to traffic patterns from intersecting facilities. Operations and safety are optimized by locating access and egress on tangent alignments.
Express Toll Lanes	• Ensure that traveler information and toll rate signage is provided in advance of the driver's decision point regarding whether to use the managed lane(s). • Barrier separation is typically preferred but often impractical due to expense and right-of-way needs. Buffer separation of two to four feet provides separation from the general purpose lanes, which can have slower travel speeds than the express toll lanes. • Managed lanes require weave zones for access and periodic widened shoulders for enforcement. While there is commonly not enough right-of-way to widen the entire length of a future managed lane, there might be certain locations along the corridor that can be preserved for such use. During the original design, avoid unnecessarily precluding these opportunities.
Truck-Only Toll Lanes	• Initial pavement design can take into account heavier design loads when a truck-only toll lane is anticipated in the future.
Reversible Lanes	• In conversion of reversible lanes from HOV to HOT operations, roadway design may require provision of adequate width at certain points in the corridor for tolling gantries and enforcement. • Enable emergency personnel to respond to incidents on a facility with limited access • Address the need for monitoring and proper deployment/closures during directional changes. • Signs and markings to indicate traffic directionality. • Provide for enforcement and tolling (if required).[37]
Contra Flow Lanes	• Movable barrier systems require the designer to identify adequate space at termini for storage of the barrier moving machine. • Provide CCTV coverage to make sure lanes are clear.
Variable Tolls	• Roadway and structure design will need to provide for future overhead signage in advance of driver decision points.
Priced Dynamic Shoulder Lanes	• Provide full-depth shoulders during normal paving operations to avoid tearing out shoulder and sub-base for future lanes. • Drainage structures and grates should be initially designed to align with wheel paths; adjustments after-the-fact can require major reconstruction. • May require slight adjustments in vertical and horizontal clearance. These can be very costly, if not prohibitive, to retrofit.

---

[36] Caltrans, "Traffic Operations Policy Directive 11-02: Managed Lane Design," 2011. Available at: http://www.dot.ca.gov/hq/traffops/systemops/hov/reference.html.

[37] FHWA, *Managed Lane Chapter for the Freeway Management and Operations Handbook*, January 2011. Available at: http://ops.fhwa.dot.gov/freewaymgmt/publications/frwy_mgmt_handbook/revision/jan2011/mgdlaneschp8/sec8.htm.

3.5 TRANSIT

During the design of a transportation facility, the transit rider must be considered just like the motorist. Transit provides the ability to increase the throughput of a facility, thereby improving overall facility operations. There are opportunities on both freeway and arterial facilities to incorporate transit operations considerations into design. One high profile application for transit on freeways is bus-on-shoulder (BOS) or bus-only shoulder. There are documented examples of BOS in California, Florida, Georgia, Maryland, Minnesota, New Jersey, Virginia, Washington, and Delaware. Minnesota is a leader in BOS operations, with more BOS lane miles than the rest of the country combined.

Minnesota DOT has developed guidelines for geometric design and signing for bus-only shoulder operations.[38] Geometric standards regarding lane widths, vertical clearance, stopping sight distance, and lateral clearance/clear zone must be considered and may not always be full design standards for retrofit applications.[39] Figure 22 shows the use of minimum 10-foot shoulders.

An FHWA report titled *Efficient Use of Highway Capacity* describes recommended spacing and design of emergency refuge areas for stalled vehicles when implementing BOS.[40]

A BOS program ensures that buses can achieve significant travel time savings by not having to enter the weave through general purpose traffic to enter or exit an interior managed lane. Some of the routes in Minneapolis experienced a 9 percent increase in ridership. A successful BOS implementation requires highway designers and transit operators to work together to implement a solution that considers ramp operations, merging, and weaving. BOS can be implemented in conjunction with managed lane strategies or ramp metering.

Another important transit operations strategy is bus rapid transit (BRT). BRT is an advanced bus system that relies on several techniques to provide faster travel times, greater reliability, and increased customer convenience over ordinary bus service. BRT offers the flexibility of buses and the efficiency of rail by operating on bus lanes or other transitways and applying advanced technologies or infrastructure such as transit signal priority and automatic vehicle location systems.[41]

Figure 22. Bus-only shoulder in Minneapolis, MN required site-specific design and operational considerations. (Source: David Gonzales, Minnesota Department of Transportation)

Florida DOT District 4 (Ft. Lauderdale) is studying transit queue jumps for use on heavily congested arterials, including the impacts of queue jumps on intersection and approaching roadway geometry. They will evaluate the traffic control devices and transit operator protocols associated with queue jumper operations and will assess their impact on other arterial traffic. They will develop a design "template" that can be utilized to identify intersections at which queue jumping should be provided and to guide the design/placement of associated traffic control devices and the design of any needed roadway modifications. The template will be systematically used in future resurfacing and other projects.

---

[38] Minnesota Department of Transportation Metro Division, *Guidelines on Shoulder Use by Buses*, 14 Jan 1997. Available at: http://www.dot.state.mn.us/metro/teamtransit/docs/bus_only_shoulder_guidelines.pdf.

[39] Jones, Greg. *Design and Maintenance: A Facilitated Roundtable Discussion*, Regional Workshop on the Use of Shoulders for Travel Lanes, FHWA, 3 May 2012.

[40] Kuhn, Beverly, *Efficient Use of Highway Capacity System*, FHWA, May 2010. Available at: http://ops.fhwa.dot.gov/publications/fhwahop10023/chap3.htm.

[41] U.S. DOT, Federal Transit Administration, Research and Technology, Bus Rapid Transit Web site. Available at: http://www.fta.dot.gov/12351_4240.html.

Transit operators understand best which strategies work best for certain corridors. Transit agencies should be engaged in design to help select the most appropriate features that allow for the maximum efficiency of the facility. Rather than retrofitting an existing freeway or arterial with these types of strategies on a case-by-case basis, they should be considered as part of an overall corridor management strategy and mainstreamed into the design process.

Table 6 identifies elements to consider during design that can impact transit operations. It also shows potential opportunities for designers to structure their roadway design (or redesign) to allow for more cost-effective implementation of transit strategies in the future. Some of these design considerations apply to multiple strategies.

Table 6. Example design considerations and opportunities for various transit strategies.

Transit Strategy	Design Considerations/Opportunities
Make HOV and HOT Lanes Accessible to Buses	• In order to optimize transit use of a managed lane facility, consider major bus routes when locating weave zones to enter and exit the lanes.
Bus Rapid Transit (BRT)	• Consider additional right-of-way to accommodate in-line stations or direct access ramps to optimize operations for stations that are adjacent to the facility. Design should allow for acceleration/deceleration lanes for buses needing to re-enter or exit lanes that may also operate as HOV lanes. For future BRT, initial roadway geometry should be designed to allow for future in-line stations and direct access ramps. • Ensure that traveler information and toll rate signage is provided in advance of the driver's decision point regarding whether to use the managed lane(s). • Pedestrian access from park-and-ride lots and circulation is critical for peak operational efficiencies and should be integral to the design process.
Designated Transit Lanes	• Initial pavement design can take into account heavier design loads when transit use is anticipated in the future.
Provide Access to Park-and-Ride Lots	• Bus access must be designed to minimize circulation and dwell time by providing direct access (dedicated ramps, priority signalization) from the adjacent highway facility. For park-and-ride lots that also function as transfer stations (bus to express bus, bus to rail), the parking side of the lot should also provide priority to buses.
Roadway DMS Used for Transit Information and Comparative Travel Times for Alternate Modes	• DMS must be placed in advance of the point on the facility where a driver can use the information to make a mode choice decision and safely weave to access a park-and-ride lot. • For future transit corridors, roadway and structure design will need to provide for overhead signage at those locations.
Park-and-Ride Space Finders	• Similar to DMS (above), these systems must be located strategically so that real-time information is transmitted to drivers at a point where it can effectively aid in decision-making regarding transit and carpool use.
Bus-on-Shoulder	• Because most bus-only shoulders are retrofit, consideration of lane delineation and signage conventions should occur in the design phase to ensure regional consistency. Bus-only shoulders are for professional drivers only, so training can be geared to operating buses within these conventions. • Provide full-depth shoulders during normal paving operations to avoid tearing out the shoulder and sub-base for future lanes. • Drainage structures and grates should be initially designed to align with wheel paths; adjustments after-the-fact can require major reconstruction.

Transit Strategy	Design Considerations/Opportunities
Arterial Bus Lanes	• Bus stop placement (near side vs. far side of intersection) has a significant impact on bus operations. Transit agencies need to be involved in the design stage as bus stop locations can depend upon the type of service (local, express). • Transit signal priority (TSP) works in conjunction with the bus stop locations to optimize express bus operations. The transit agency should have input in the design of the facility and the TSP software programming. • Real-time arrival displays aid riders in selecting bus routes. The design needs to provide for electrical and communications connections. • Where space permits, queue jump lanes can be used at signalized intersections in conjunction with TSP to reduce dwell time at stops. Adding queue jump lanes requires transit agency input in the design process.

3.6 WORK ZONE MANAGEMENT

Managing traffic during construction is necessary to minimize traffic delays, maintain or improve motorist and worker safety, complete roadwork in a timely manner, and maintain access for businesses and residents. Work zone traffic management strategies should be identified based on project constraints, construction phasing/staging plan, type of work zone, and anticipated work zone impacts.[42]

Agencies should consider performance-based maintenance of traffic requirements, such as maximum allowable delay, rather than geometric or time of day restrictions. This approach allows greater creativity and innovation by contractors, which may result in both cost savings to the agency and time savings to motorists.

A transportation management plan (TMP) is a successful approach to identifying transportation management strategies and describing how they will be used to manage the work zone impacts of a project. The FHWA publication *Developing and Implementing Transportation Management Plans for Work Zones* defines planning and design considerations for work zone management. Throughout the development of a TMP, designers and operational stakeholders have the opportunity to consider the impacts of their work zones and to identify strategies to improve work zone performance.[43] The TMP is primarily intended for managing traffic during a construction project. However, some of the elements of the TMP, particularly ITS improvements, could remain in place to aid ongoing operations. Additionally, the cross-functional and interagency relationships formed during the development and use of the TMP should be continued after the project to promote a coordinated approach to operating the facility.

The inclusion of work zone management and operations should be identified during needs development and preliminary engineering so that strategies can be implemented prior to the start of major construction activities if needed. In addition, the transportation facility should be designed with construction and post-construction maintenance of traffic activities in mind. Designers must consider how the facility will be constructed in a manner that provides a safe working environment and minimizes the impact on the operation of the facility. This may require consideration of construction methods and staging.

Traffic capacity and shoulder/pullout areas are often restricted in work zones. Prompt detection and clearance of traffic incidents in work zones can help reduce secondary crashes and delay. Preparing a work zone TIM plan and using strategies that improve detection, verification, response, and clearance of crashes, mechanical failures, and other incidents in work zones and on detour routes can benefit safety and mobility. Specific strategies are identified in FHWA's, *Traffic Incident Management in Construction and Maintenance Work Zones*.[44]

---

[42] FHWA, "Work Zone Traffic Management," *Work Zone Mobility and Safety Program*. Available at: http://www.ops.fhwa.dot.gov/wz/traffic_mgmt/index.htm.

[43] FHWA, *Developing and Implementing Transportation Management Plans for Work Zones*, December 2005, FHWA-HOP-05-066. Available at: http://www.ops.fhwa.dot.gov/wz/resources/publications/trans_mgmt_plans/index.htm.

[44] FHWA, *Traffic Incident Management in Construction and Maintenance Zones*, FHWA-HOP-08-056x, 2008. Available at: http://ops.fhwa.dot.gov/publications/fhwahop08056x/execsum.htm.

Additional Work Zone Strategies

- Automated work zone information systems and automated work zone enforcement systems (if legislation allows).
- Traffic management systems such as dynamic lane merge systems and speed management systems.
- Capacity enhancements such as contraflow lanes and express lanes (i.e., lanes in which high levels of congestion are managed typically by varying the toll price by time of day or level of traffic).
- Improvements to detour routes.
- Incident pull-off areas, incident staging areas and investigation sites.
- Incident turnarounds and access gates.
- Evaluate need for temporary traffic signals and integrate them into the existing system so they can be timed appropriately.

3.7 TRAFFIC INCIDENT MANAGEMENT

Traffic incident management (TIM) practitioners become well aware of shortfalls in operational provisions when it affects their ability to respond to incidents safely and efficiently. There are several ways that designers can ensure that the needs of this end-user group are considered and included in the final design of a project. The National Unified Goal (NUG) for TIM is a foundational element of a well-developed TIM Program and provides a valuable opportunity to link program decisions to physical design. Table 7 identifies elements to consider during design that can address various NUG strategies.[45]

Table 7. Example design considerations and opportunities for NUG TIM strategies.

NUG Strategy[46]	Design Considerations/Opportunities
Strategy 1 – Partnerships and Programs. Partners should work together to develop and promote public awareness and education about roadway and incident safety.	• Locate DMS where safety concerns exist to raise awareness. • Partnerships that promote laws in support of TIM (e.g., Move-It, Move-Over, Hold Harmless, Quick Clearance).
Strategy 4 – Technology. Partners should work together for rapid and coordinated implementation of beneficial new technologies.	• Employ ITS standards that promote consistency and interoperability. • Include a system interoperability plan for agencies responsible for detecting and verifying incidents.
Strategy 6 – Awareness and Education. Broad partnerships to promote public awareness of their role in safe roadways.	• Include the provision to hold coordination meetings as part of the design phase that focuses on emergency services.

---

[45] National Traffic Incident Management Coalition, National Unified Goal for Traffic Incident Management Detailed Explanation. Available at: http://ntimc.transportation.org/Documents/NUG-4pp_11-14-07.pdf.
[46] Ibid.

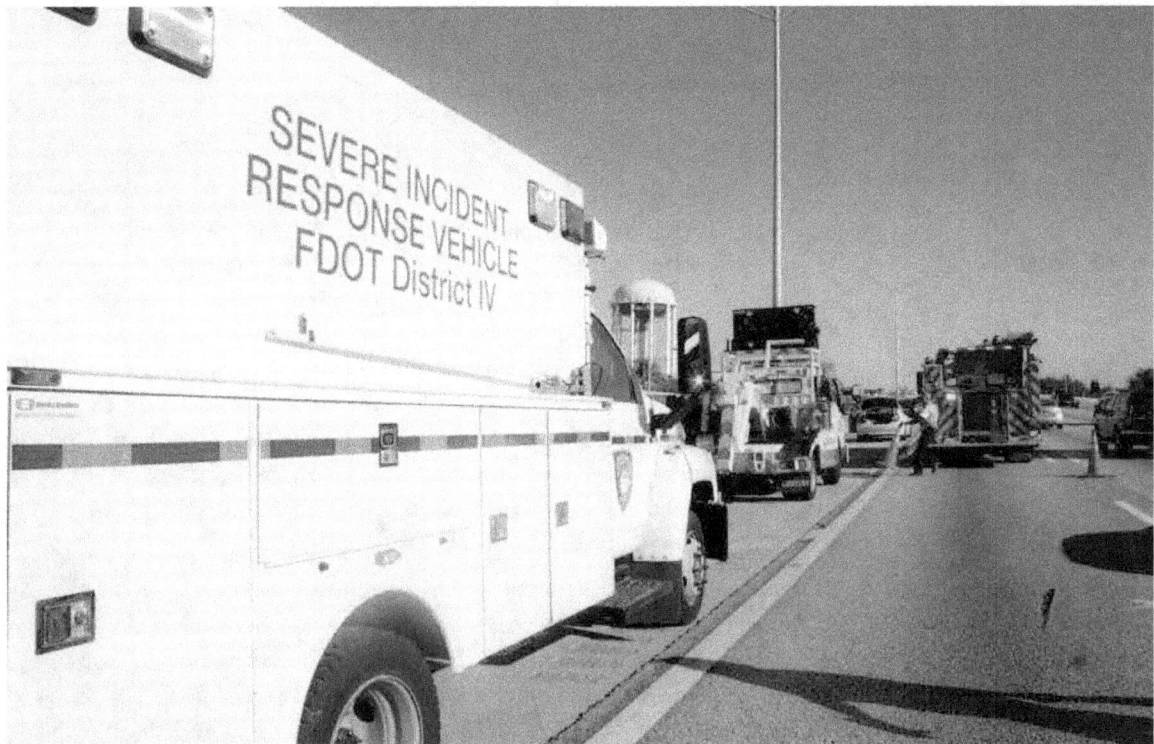

Figure 23. One of the design considerations for TIM is shoulder width for accommodating response vehicles.
(Source: Florida Department of Transportation)

NUG Strategy	Design Considerations/Opportunities
Strategy 7 – Recommended Practices for Responder Safety. Recommended practices for traffic control at incident scenes should be developed and widely published and adopted.	• Expand maintenance and operations of traffic plans to include considerations for responder safety such as temporary barrier placement, temporary shoulder width, and others.
Strategy 11 – Response and Clearance Time Goals. Partners should commit to achievement of goals for response and clearance times.	• Provide median breaks and crash investigation/motorist information exchange sites (can include "fender bender" signage to direct non-injury accident vehicles out of traffic). • Provide static signs directing responders to investigation sites, including 1/10 mile markings and a system to identify locations on ramps within complex interchanges. • CCTV should be designed and constructed in high-crash locations to improve detection and verification time.
Strategy 17 – Prompt, Reliable Traveler Information Systems. Partners should encourage development of more prompt and reliable traveler information systems that will enable drivers to make travel decisions.	• Incorporate incident information into pre-trip (e.g., 511) and en-route traveler information services and alerts.

Permanent Median Crossovers

At approaches to major bridges or freeway segments where there are long distances between exits, designers should consider converting construction detour crossovers to permanent cross over facilities to accommodate detours for incident management. The crossover should have proper treatments, such as delineators, to protect against wrong-way use.

During the design phase, the project team should seek input on roadside safety from emergency responders or a TIM team at important milestones, such as the transition from preliminary engineering to final design. Input from responders on roadside features such as noise walls, median barriers, and ITS device locations should be considered a priority in the design process. If a local TIM team does not exist where the project will be located, the project team should establish one with the goal of creating a framework that will ensure continued TIM team existence after the project is complete. During construction, the TIM team should be engaged to ensure that both constructability and emergency response risks are balanced.

It is important to develop a good rapport with emergency responders during a construction project. An effective way to make the best use of their time and to gain valuable insights into their operational needs is to conduct a table-top exercise that includes the proposed design. After the design plans have reached a level that makes it clear what will change from the existing situation, the design team should gather the local TIM team members or establish the TIM team and conduct a table-top exercise to "test" the design for operations. In addition to agency design personnel, the team should include maintenance staff and emergency responders. It is suggested that at least three scenarios be included during this session to generate discussion:

1. A crash and subsequent release of hazardous materials.
2. A full directional blockage during construction when access is reduced.
3. A full directional blockage for the final condition.

In addition to documenting the needs of emergency responders in each of these scenarios, there should also be discussion about how the response to these scenarios differs if construction workers are present at the incident site. The response to less severe events should also be covered.

3.8 SECURITY

Transportation agencies need to deploy appropriate risk reduction methods to minimize or eliminate identified vulnerabilities in their system, and designers need to consider if countermeasures are appropriate for their particular project. *NCHRP Report 525 - Surface Transportation Security* discusses many of the tools and countermeasures that should be considered in the design phase as a means to improve the security of critical infrastructure and facilities, information systems, and other areas.[47] Physical security countermeasures that should be considered by a designer may include signs; emergency telephones, duress alarms, and assistance stations; key controls and locks; protective barriers; protective lighting; alarm and intrusion detection systems; electronic access control systems; and surveillance systems and monitoring.

Agencies must conduct threat and hazard analyses for use in prioritizing the most important roads and infrastructure. Controlling access to critical components, providing standoff from critical components, eliminating single point of failure construction, and ensuring that surveillance systems are tied directly into response units are the best strategies to deter or prevent terrorist or criminal acts. Many of these strategies are very costly and must be considered in the scoping phase. Even though making these decisions is beyond the authority of the individual designer, there are related elements that can be considered in the design phase.

Designers should contact internal and stakeholder security and emergency management officials to develop security and emergency management requirements. This coordination can prevent issues such as designing and building a structure for standard loads then retrospectively learning that it is a critical primary route that must be designed for moving

---

[47] NCHRP Report 525, *Surface Transportation Security Volume 1, Responding to Threats: A Field Personnel Manual*, Transportation Research Board/National Academies of Science, 2004.
Available at: http://onlinepubs.trb.org/onlinepubs/nchrp/nchrp_rpt_525v1.pdf.

heavy equipment into an area during an emergency. Security and emergency management planning and designing takes a community of people drawn from law enforcement/security, fire and emergency medical services, emergency management, occupational safety, and highway/transportation organizations.

Table 8 identifies elements to consider during design that can impact infrastructure security. Transportation agencies must examine the threats against infrastructure and identify the most useful means to reduce the vulnerabilities associated with those threats to acceptable levels. Often less costly but more effective solutions are available that the agency can select to meet security requirements. In making these choices, designers can benefit from an analysis that compares one countermeasure against another based on protection provided, cost, and effort required.

Table 8. Example design considerations and opportunities for various security strategies.

Security Strategy	Design Considerations/Opportunities[48]
Monitoring Systems	• Place CCTV systems in areas of high interest, such as bridges and tunnels. • Include alarms on access doors and equipment cabinets.
Preventative Infrastructure Design	• Place barriers or gates at on/off-ramps to close a road during an emergency. • Increase the "stand-off" or buffer distance around bridge abutments or tunnel entrances. • Install dolphins or fender systems around bridge supports in navigable waterways to protect them from intentional accidents or impacts. • Incorporate "web walls" between bridge piers to strengthen them to better resist damage from vehicle wrecks or train derailments.

3.9 FREIGHT OPERATIONS

Freight operations are an important consideration with respect to improving mobility and productivity. Improved operation can benefit the freight industry through:

- Immediate cost reductions to carriers and shippers, including gains to shippers from reduced transit times and increased reliability, resulting in decreased cost of raw materials and finished goods.

- Reorganization-effect gains from improvements in logistics. The quantity of firms' outputs changes, but quality of output does not.

- Gains from additional reorganization effects such as improved products or new products.[49]

Figure 24. Weigh-in-motion station. (Source: Tennessee Department of Transportation)

Additionally, improving freight operations enhances the safety and efficiency of the transportation system for all users by lessening the impact of freight movements on the general public and vice versa. Virginia DOT has been focused on improvements geared toward truck traffic along Interstate 81 (its most heavily traveled truck route) for years. Improvements include interchange redesign, truck climbing lanes, ITS improvements, and ramp extensions. During design, however, consideration of freight must extend beyond the geometric considerations associated with commercial vehicles to include operational elements that support enforcement and hours-of-service requirements, as well as elements to improve safety and overall efficiency.

---

[48] FHWA, *Considering Security and Emergency Management in the Planning of Transportation Projects*, May 2012. Available at: http://www.planning.dot.gov/documents/ConsideringSecurityAndEM.pdf.

[49] FHWA, *Freight Benefit/Cost Study: Compilation of the Literature*. February 2001. Available at: http://ops.fhwa.dot.gov/freight/documents/freight_bca_study.pdf.

The table below illustrates specific actions that designers can take to enhance freight operations. In some cases, where significant infrastructure improvements are involved, strategies must be initially considered in the scoping/planning phase. However, designers can optimize the effectiveness of these strategies through use of the specific design considerations.

Table 9. Example design considerations and opportunities for various freight strategies.

Freight Operations Strategies	Design Considerations/Opportunities
Consider Trucks as a Discrete Mode with Different Characteristics than Passenger Vehicles	• Consider turning radii, lane widths, ramp geometry, acceleration/deceleration lanes, directed signing. • Need for urban loading zones, delivery windows, signal timing, turning lane lengths. • Match truck routes with appropriate infrastructure, considering height and weight constraints. • Make freight operational improvements part of the total system to avoid downstream effects.
Improve Size and Weight Enforcement to Extend Infrastructure Life	• Mitigate noise, visual, and air pollution by enforcing regulations and decreasing congestion. • Include weigh-in-motion stations to improve enforcement and reduce delays. • Embrace automated inspection technology. • Utilize commercial vehicle information systems and networks and electronic credentialing. • Ensure appropriate truck route, clearance, and weight limit signing system-wide.
Consider Infrastructure and Systems that Improve Driver and Vehicle Safety	• Provide rest areas and services for long-haul drivers. • Deploy "smart" truck parking systems that provide information on available parking spaces to upstream truck drivers. • Deploy over-height vehicle detection systems and comprehensive restrictions signing where over-height crash rates are high. • Deploy truck escape ramps on severe downgrades. Designers can work with the trucking industry and operations staff to identify locations and designs appropriate for each specific steep grade. • Implement truck restrictions such as "no passing, right lane only." • Accommodate appropriate shoulder and travel lane widths on primary and secondary roadways. • Include slow moving vehicle lanes (upgrade/downgrade). Designers should consider truck acceleration/deceleration and other characteristics in locating termini of these lanes.

3.10 MAINTENANCE

Maintenance of a roadway can have a major effect on operations. Maintenance personnel have a variety of issues to deal with; from mowing operations in the summer, to snow plowing operations in the winter, to maintenance of roadside devices, they are constantly working to keep roadway networks operating. Taking into consideration certain aspects of the design of the roadway and devices can reduce the impacts of maintenance operations. For example, inadequate shoulder widths may require maintenance personnel to shut down a lane to perform their duties.

Figure 25. Grade warning and runaway truck ramp location sign for trucks. (Source: Tennessee Department of Transportation)

Including maintenance personnel in the design process can help designers identify many maintenance issues that they may not be aware of. There are two ways this can be achieved. First, during the design phase, the project team should invite maintenance personnel to design meetings where they can provide input on design aspects. Input from maintenance personnel on roadside features such as noise walls, median barriers, and ITS device locations should be considered a priority in the design process.

Second, agencies should include processes or checklists in design manuals to obtain sign-off on plans from their maintenance division. Maintenance personnel should comment on issues that relate to snow plowing operations, barrier selection and placement, impact attenuator selection, signal systems and ITS infrastructure, landscaping, median crossovers/turnarounds, shoulder width, and culvert treatments, among others.

Figure 26. The utilization of two types of median treatments may impact mowing activities. (Source: SAIC)

Table 10 identifies elements to consider during design that can impact maintenance operations. It also shows potential opportunities for designers to structure their roadway design (or redesign) to allow for more cost-effective implementation of maintenance strategies in the future. Some of these design considerations would apply to multiple strategies.

Table 10. Example design considerations and opportunities for various maintenance strategies.

Maintenance Strategy	Design Considerations/Opportunities
On-going Routine and Preventive Maintenance	• Incorporate areas for maintenance equipment and personnel to safely address immediate and small areas of pavement repair (e.g., potholes). • Provide portable DMS to alert drivers to moving operations. • Locate lighting to minimize knockdowns. • Provide fall protection elements on bridges for maintenance personnel. • Provide maintenance access for stormwater management facilities.
Managing Preventive Maintenance Impacts (e.g., shoulder and lane widths)	• Ensure that improved and new shoulders are wide enough to accommodate typical operations and maintenance vehicles without encroaching on travel lanes. • Provide areas behind guardrail for maintenance personnel to work or pull over their vehicles and equipment. • Provide brackets for sign structure lighting that allows it to be swung to the side of the road so that lane closures are not necessary with working on the lighting. • Consider maintenance needs during crash cushion selection.
Roadside Equipment (e.g., DMS, CCTV, other ITS, signs, lighting)	• Consider placing ITS devices near bridges to prevent the need for a lane closure to maintain the device. • Provide a proper workspace around roadside equipment for an operator/repair team to access the equipment. • Consider providing stone drives to ITS devices in median that have to be protected by guardrail. • Provide locations for bucket trucks to park to access ITS devices. • Design catwalks and signs that can be rotated for ease of access without closing lanes to minimize traffic disruptions.
Mowing Operations	• Place center median guardrail so that it allows mowing without having to place any portion of the mower on the pavement. • Limit steep slopes so that mowing can occur without specialized mowers.
Plowing Operations	• Flare energy absorbing terminals away from the lane of travel.

4 A Way Forward

Over the past decade, transportation agencies have become increasingly focused on providing the greatest level of mobility, safety, and security with their roadway infrastructure investments. This is due to a number of factors, such as increased demand, limited space for roadway expansion, and funding and environmental constraints. It requires the use of M&O strategies, which maximize the use of traditional infrastructure through managing transportation smartly with traveler information, TIM, managed lanes, and other approaches. The use of M&O strategies requires infrastructure—both roadway and ITS—that is designed with operations in mind. Otherwise, the roadway creates a range of impediments to using these strategies, and either costly modifications must be made or the strategy is performed inefficiently, if at all. Many agencies have struggled with this problem and several of them are proactively addressing it through policies, guidance, training, or increased collaboration. This primer highlights examples from agencies making strides in designing for operations, such as the Pennsylvania DOT, Caltrans, and Portland Metro. These examples of emerging practices are intended to motivate, inspire, and offer additional resources in your pursuit of your agency's unique approach to designing for operations.

While each transportation agency, State, or region has unique circumstances that will dictate an individual approach to accounting for M&O strategies during project development and design, this primer introduces key elements that agencies will need in developing a way forward. Agencies will need policies and an organizational structure that prioritizes operations in infrastructure design and institutionalizes the process of designing for operations. For example, the Missouri DOT and Caltrans have policies of cross-functional collaboration (including traffic operations) during the infrastructure design process. Several transportation agencies are elevating operations in their organizational structures by creating high-level operations departments or integrating an M&O program throughout each organizational division.

An effective designing for operations approach will also require direction for the design or project development team on what to consider. This primer recommends closely linking the design process to the planning process such that the collaborative decisions made during the planning process guide considerations. Any objectives and performance measures for transportation system operations should be used in evaluating design alternatives so that the infrastructure supports the area's ability to reach its operations objectives. The infrastructure should be designed to support applicable M&O strategies that have been selected through the planning process or through the development of an operations strategic plan or regional ITS architecture. What to consider during the design process should also be informed by agency policy, internal and external operations professionals, and input from stakeholders in areas such as transit agencies, pedestrian and bicycle advocates, commercial vehicle operators, and other important infrastructure user groups. Section 3 of this primer was built with the contributions of several operations and design experts across the United States with the purpose of providing you with an initial list of tangible design considerations that help support or are required in the use of a number of popular M&O strategies. This section can be modified and distributed among your project design team as an initial step in tailoring a designing for operations approach that works for your organization.

Figure 27. Aerial view of truck weigh station. (Source: Florida DOT)

Finally, the design team needs the knowledge of how to design roadway infrastructure and deploy ITS to enable M&O strategies. This knowledge is most often provided through formal design guidance such as Caltrans' *Ramp Metering Design Manual*, training, and cross-functional collaboration where operations experts and roadway designers work together to develop project-specific designs treatments.

On a national level, this primer represents the first step in promoting the frequent and systematic consideration of M&O strategies across each stage of the design process. Soon, the SHRP 2 L07 project, *Evaluation of Cost-Effectiveness of Highway Design Features*,[50] will provide another tool to transportation agencies to help evaluate the operational and safety impacts of multiple design treatments. Future efforts to advance designing for operations may include national guidelines or standards similar to those for geometric design found in AASHTO's *A Policy on Geometric Design of Highways and Streets*, also known "The Green Book."[51] The FHWA will continue to support the widespread adoption of designing for operations practices across the United States to support agencies in maximizing the operational benefits from their roadway investments.

---

[50] Transportation Research Board, SHRP 2 L07, project description for "Evaluation of Cost Effectiveness of Highway Design Features." Available at: http://apps.trb.org/cmsfeed/TRBNetProjectDisplay.asp?ProjectID=2181.

[51] AASHTO, *A Policy on Geometric Design of Highways and Streets*, 2011. Available for purchase at: https://bookstore.transportation.org/collection_detail.aspx?ID=110.

www.ingramcontent.com/pod-product-compliance
Lightning Source LLC
Chambersburg PA
CBHW081908170526
45167CB00007B/3197